ACPL ITEM DISCARDED

665.
Pear
Guid...
reference literature

**DO NOT REMOVE
CARDS FROM POCKET**

ALLEN COUNTY PUBLIC LIBRARY
FORT WAYNE, INDIANA 46802

You may return this book to any agency, branch,
or bookmobile of the Allen County Public Library.

DEMCO

Guide to the
Petroleum Reference Literature

Reference Sources in Science and Technology Series

Guide to the Literature of Pharmacy and the Pharmaceutical Sciences. By Theodora Andrews.

Guide to the Petroleum Reference Literature. By Barbara C. Pearson and Katherine B. Ellwood. Foreword by Arthur R. Green.

Guide to the Petroleum Reference Literature

Barbara C. Pearson
and
Katherine B. Ellwood

Foreword by
Arthur R. Green

1987

Libraries Unlimited, Inc.   **Littleton, Colorado**

Copyright © 1987 Libraries Unlimited, Inc.
All Rights Reserved
Printed in the United States of America

No part of this publication may be reproduced, stored in a retrieval system, or transmitted, in any form or by any means, electronic, mechanical, photocopying, recording, or otherwise, without the prior written permission of the publisher.

LIBRARIES UNLIMITED, INC.
P.O. Box 263
Littleton, Colorado 80160-0263

Library of Congress Cataloging-in-Publication Data

Pearson, Barbara C., 1950-
   Guide to the petroleum reference literature.

   (Reference sources in science and technology series)
   Includes index.
   1. Petroleum--Bibliography. 2. Petroleum--Information services. I. Ellwood, Katherine B., 1957- . II. Title. III. Series.
Z6972.P33   1987   [TN870]   016.6655   86-21434
ISBN 0-87287-473-7

Libraries Unlimited books are bound with Type II nonwoven material that meets and exceeds National Association of State Textbook Administrators' Type II nonwoven material specifications Class A through E.

Contents

Foreword vii

Preface ix

Acknowledgments xi

1 – Guides to the Literature and Bibliographies 1

2 – Indexing and Abstracting Services 5

3 – Dictionaries 10
    General Works 11
    Specialized 12
    Multilingual 14

4 – Encyclopedias and Yearbooks 16
    General Works 17
    Europe 19
    Great Britain 20
    Japan 20
    Middle East 20
    United States 21

5 – Handbooks, Manuals, and Basic Texts 22
    General Works 23
    Business 26

5 — Handbooks, Manuals, and Basic Texts (*continued*)
    Drilling 27
    Exploration 30
    Legal 31
    Offshore 33
    Pipeline 34
    Production 36
    Refining 37
    Style Manuals 38

6 — Directories 39
    Individuals 40
    Companies and Other Organizations 42
        International 42
        United States 45
        Canada 51
        Europe 52
        Latin America 53
        Middle East 54
        Southeast Asia 54
    Products and Equipment 54

7 — Statistical Sources 59

8 — Databases 65
    Bibliographic 66
    Data/Textual 81
        Full Text 91
    Data Only 93

9 — Periodicals 130

10 — Professional and Trade Associations 145

11 — Publishers 158

Author/Title Index 167

Subject Index 179

Foreword

Civilizations, nations, and corporations rise, flourish and fall in relation to their utilization of two kinds of resources—natural and human. The human mind with its creative ability has made it possible to find, extract, produce, and transform natural resources into energy and products.

We are living through a major turning point in the evolution of mankind which allows us to utilize our human resources as never before. We have entered the age of information and communication. Alvin Toffler refers to it as the third wave of civilization. In the first wave, the agricultural revolution made the land more productive and changed the lives of mankind. The second wave of industrial revolution utilized new technology to enhance our standard of living. Advanced technology has now spawned a swell that is breaking into a new wave. The ability to recall, communicate, and integrate knowledge at high speed is propelling us into a third wave of information, communication and knowledge.

Information is power! From it we can derive knowledge. From knowledge we make wise business decisions and advance our technology. Concepts and judgments cannot be better than the information on which they are based.

The cumulative knowledge of the human race, and the ability to file and retrieve memories, is what has set us apart. We are able to learn from those who have gone before us. Man's early cumulative knowledge was held in the minds of individuals and passed on by tribal elders. Next, it was put down on paper and filed in libraries, museums, and the offices of government and industry. Now, with computers to store almost unlimited bits of information, we can recall vast amounts of knowledge and sift, integrate, and recombine this information in new ways. We can recall beyond the storage capacity of our individual minds and extract information and knowledge from society's cumulative memory. Thus, we can "think" about things we have never known or experienced.

Information supplies the resource for this creative and innovative thinking. Unlike our natural resources, this is a resource that can't be expended—indeed it is growing at an accelerated pace. This ability to store and retrieve information from beyond our individual minds, is in turn increasingly stimulating our brains, and the implications for our civilization, our nation, the business world, and indeed the individual, are profound.

John Adams would feel comfortable in this new wave of information, for he wrote, "Let us cherish, therefore, the means of knowledge. Let us dare to read, think, speak, and write ... let every sluice of knowledge be opened and set a-flowing."

It has been said many times, in many ways, that there are two kinds of knowledge—the knowledge one has of a subject itself and the knowledge to find sources of new information about the subject. There is nothing so captivating as discovering new information that bears directly on a problem to be solved or a decision to be made.

Those involved in finding, producing, and refining energy and raw materials now have a new wave of information breaking before them. This book is a product, an indication, and itself a catalyst of this information "wave." It is a guide to those who wish to enter the new and exciting era of information, communication, and knowledge.

Arthur R. Green
Research Scientist
Exxon Production Research Company

Preface

Intense competition exists to develop more efficient and economical ways to find and recover oil. The literature that emerges as a result of these efforts provides the building blocks for petroleum technology. Those who are able effectively to use this literature often gain the competitive edge. Reference sources traditionally provide pathways into the literature. They may provide the actual answer to a question or they may lead the reader to additional appropriate sources.

The petroleum industry draws from many disciplines in the course of its day-to-day operations. An understanding of geology, geophysics, and geochemistry is necessary in the exploration for oil and natural gas. Once the oil and gas have been located, production and refining operations depend on the successful integration of chemistry, physics, and engineering.

Many guides to the literature have been written for geology, chemistry, physics, and engineering. This guide, the first to focus on petroleum sources, also includes some major works in these related disciplines as they apply to the petroleum industry. Petroleum reference sources include bibliographies, abstracts, indexes, dictionaries, encyclopedias, yearbooks, directories, handbooks, databases, statistical sources, and periodicals written specifically for the petroleum industry. Although special attention has been given to current reference materials, some of the older, "classic" sources have also been included.

This guide includes a selected list of reference tools chosen from collections in major petroleum-company and university libraries, such as the University of Texas at Austin Library, the Texas A & M University Library, and the Rice University Library. Materials sent from scientific publishers for our review, and sources suggested by members of the Special Libraries Association and Geoscience Information Society, were also considered for inclusion. Materials selected for inclusion are international in scope, primarily in English, and for the most part have been published since 1978. All

x / PREFACE

reference books included were evaluated personally by the authors. Descriptions of databases, however, are based on information supplied by producers and vendors.

Eleven chapters present detailed discussions of guides and bibliographies; indexing and abstracting services; dictionaries; encyclopedias and yearbooks; handbooks, manuals, and basic texts; directories; statistical sources; databases; major periodicals; professional and trade associations; and key publishers. Entries in the first seven chapters include bibliographic information and a concise description of the arrangement, content, and use of the reference work. Every effort has been made to provide ISBN, ISSN, and/or LC numbers for each entry to help the reader identify these references. Occasionally, other identifying numbers have been added. Prices, when available, have also been included as a general indication of the cost of the materials at the time of this guide's publication. Entries in chapter 8 include database producer, format, update frequency, vendor, availability, etc., as well as a description of content. Chapter 9 provides name, address, and ordering information for major periodicals. Entries in the final two chapters provide current addresses for professional/ trade associations and publishers. Some annotations include descriptive statements from the introductory material of the work discussed.

Guide to the Petroleum Reference Literature is intended to help students, information specialists, librarians, and professionals associated with the petroleum industry. Reference books are never static! New information and sources become available almost daily. The authors of this guide welcome suggestions from the reader on additional sources to be included in future editions.

Acknowledgments

We want to thank all those who have been so generous with their time and support. Specifically, we would like to thank Dan Ellwood, Joe Pearson, Linda Neubauer, Laura Rustin, and the staffs of the Exxon Production Research Company Information Center and the Texaco Houston Research Center Library. We couldn't have done it without you!

1 Guides to the Literature and Bibliographies

Bibliographies and guides to the literature provide a valuable starting point in trying to locate information in a given subject area. Although several bibliographies containing information on petroleum have been published over the years, until now there has not been a guide solely dedicated to the petroleum industry. This chapter discusses a few subject-specific guides—*Energy Information Guide* (entry 4), *Geologic Reference Sources* (entry 5), and *Guide to the Energy Industries* (entry 6)—that contain petroleum-related information. *Science and Engineering Literature* (entry 8) is a good example of a general science guide which is also useful in locating petroleum information. Guides exist for other major scientific disciplines (chemistry, physics, engineering, mathematics) that have an impact upon the petroleum sciences. These sources can be identified in reference tools such as *Walford's Guide to Reference Materials*[1] or *Guide to Reference Books*.[2]

Online database searching has improved the capability to rapidly produce current printed bibliographies on a specific topic. This capability has had an important impact on the number of petroleum bibliographies published in recent years. The *Bibliography and Index of Geology* is an example of one petroleum-related bibliography which is currently available in both printed and online formats (see entries 1 and 243). Chapter 8 includes descriptions of other petroleum databases which may also be used to produce excellent petroleum bibliographies.

---

[1]A. J. Walford, ed., *Walford's Guide to Reference Materials*, 4th ed. (London: The Library Association, 1980). ISBN 0-85365-611-8.

[2]Eugene P. Sheehy, comp., *Guide to Reference Books*, 9th ed. (Chicago: American Library Association, 1976). LC 76-11751. ISBN 0-8389-0205-7.

2 / 1 – GUIDES TO THE LITERATURE AND BIBLIOGRAPHIES

1. **Bibliography and Index of Geology.** Alexandria, Va.: American Geological Institute, 1969- . monthly. $975.00/yr. LC 75-642313. ISSN 0098-2784.

This bibliography is often referred to as the "geology bible" by students, professors, researchers, and librarians. It offers comprehensive, worldwide coverage of the earth science literature. In 1969, the *Bibliography and Index of Geology* succeeded the *Bibliography and Index of North American Geology* (1785-1970) and the *Bibliography and Index of Geology Exclusive of North America* (1933-1968). The *Bibliography and Index of Geology* (*BIG*) is issued monthly and is cumulated once a year. Each monthly issue has four subdivisions: serials, field of interest, subject index, and author index. Approximately 70 percent of the references in the *BIG* are from serial publications. Each entry includes author, full title, publication source, page numbers, and date. Some entries contain additional information in notes following the bibliographic citation. The online version of the *BIG* is called GeoRef (see entry 243).

2. **Current Bibliography of Offshore Technology and Offshore Literature Classification.** 2nd ed. By Arnold Myers. Hertfordshire, England: ASR Marketing, 1984. 167p. indexes. $80.00. ISBN 0-906528-02-X.

The intent of this bibliography is to "list the material that should be purchased by a library setting up an offshore technology section in order to provide a thorough basic reference collection." The importance of conference proceedings is stressed and a listing of conferences and seminars with published proceedings is included. The bibliography is divided into three broad categories: general, oil and gas, and marine technology. Each entry supplies author, title, and bibliographic citation. A detailed publishers' address list is provided.

3. **Energy Bibliography and Index.** Houston, Tex.: Gulf Publishing Co., 1978-82. 5v. $375.00/vol. LC 77-93150.

This index is intended to be a comprehensive guide to energy-related materials located in the Texas A & M University Library. "The EBI includes entries for books, serials, periodical titles, government documents, selected technical reports, conference proceedings and symposia, maps, pamphlets and brochures." Journal articles are not included. Each entry in the bibliography is annotated. Some of the petroleum-related topics covered by the index are petroleum demand, petroleum deposits, environmental aspects of petroleum, and petroleum refineries. The acronym for the online version of this index is EBIB (see entry 239).

4. **Energy Information Guide.** By R. David Weber. Santa Barbara, Calif.: ABC-Clio, 1982. 3v. $39.95/vol. 1; $39.95/vol. 2; $44.50/vol. 3. LC 82-8729. ISBN 0-87436-317-9 (vol. 1); 0-87436-341-X (vol. 2); 0-87436-343-8 (vol. 3); 0-87436-388-8 (set).

This guide represents a monumental effort to provide students, information specialists, researchers, and general readers with access to current, as well as historic, reference sources in the field of energy. The three volumes are divided by main subjects. Chapters in volume 1 include "General Energy Sources," "Conservation," "Alternative Energy Sources," "Solar Energy," "Wind Power," "Geothermal Energy," and "Other Energy Alternatives." Volume 2 concentrates on electric power and nuclear power. Volume 3 includes chapters entitled "General Sources for Fossil Fuels," "Petroleum and Natural Gas," and "Coal." Dictionaries, encyclopedias, handbooks, manuals, yearbooks, directories, statistical sources, indexes, abstracts, bibliographies, and databases are annotated for each subject area. Annotations are generally lengthy and often provide historical information about the source. Entries are numbered

1 – GUIDES TO THE LITERATURE AND BIBLIOGRAPHIES / 3

sequentially throughout the three volumes. Each volume of the guide has four indexes: author, subject, title, and document number. The indexes are cumulative.

5. **Geologic Reference Sources: A Subject and Regional Bibliography of Publications and Maps in the Geological Sciences.** 2nd ed. By Dederick C. Ward, Marjorie W. Wheeler, and Robert A. Bier, Jr. Metuchen, N.J.: Scarecrow Press, 1981. 560p. indexes. $30.00. LC 81-4770. ISBN 0-8108-1428-5.

This annotated guide to geoscience reference literature is divided into three sections. Section 1, "General Information Sources," is arranged by type of source (e.g., directories, encyclopedias, and bibliographies). Section 2, "Subject Section," is topically arranged (e.g., oceanography, geochemistry, and paleontology). Section 3, "Regional Section," divides sources geographically. Entries are consecutively numbered throughout the three sections. A separate subject and geographic index is provided. This second edition includes a current awareness section and a discussion of new ways to access geological information through online bibliographic retrieval services. This source will be useful for undergraduate and graduate students, information specialists, and researchers.

6. **Guide to the Energy Industries.** Cambridge, Mass.: Ballinger Publishing Co., 1983. 328p. indexes. $60.00. ISBN 0-88410-919-4.

This four-part guide emphasizes marketing and financial data for seven energy sources: coal, energy alternatives, hydroelectric power, natural gas, nuclear energy, petroleum, and solar. Although directed to economists and policy analysts, this source is also quite useful to information specialists and researchers. It contains more than twenty-nine hundred entries. Entries include market research reports, investment banking reports, forecasts, economic studies, sources of statistics, handbooks, directories, newsletters, and numeric databases. Part 1 includes consecutively numbered, annotated references, arranged alphabetically by type of energy source. Part 2 is divided into two sections. The first is an alphabetical listing of energy-related publishers. The second is arranged by type of document, then alphabetically by name of publisher. Part 3 contains the subject and SIC code indexes, while part 4 is the title index. Most items included were published between 1980 and 1982.

7. **Offshore Petroleum Engineering.** By Marjorie Chryssostomidis. New York: Nichols Publishing Co., 1978. 367p. index. $50.00. LC 78-16114. ISBN 0-89397-045-X.

Both offshore engineering and offshore development are covered in this partially annotated bibliography. It is useful to people new to the offshore engineering industry, those already familiar with this literature and anyone involved in collection development in this field. Arrangement is by broad subject area, then by more specific subsections. Within each subsection, entries are grouped according to format (i.e., conferences, reports, books, etc.). There is a strong emphasis on sources other than journal articles and conference papers. Order information is given in a section titled "How to Locate and Obtain Publications" and a directory of publishers and other organizations. Three indexes are provided: a permuted topic index, an author index, and a title index.

8. **Science and Engineering Literature: A Guide to Reference Sources.** 3rd ed. By H. Robert Malinowsky and Jeanne M. Richardson. Littleton, Colo.: Libraries Unlimited, Inc., 1980. 342p. index. $33.00. LC 80-21290. ISBN 0-87287-230-0.

4 / 1 – GUIDES TO THE LITERATURE AND BIBLIOGRAPHIES

This guide is an annotated introduction to reference sources for the science and engineering literature. It is useful to students as well as reference librarians. The guide is divided into fourteen chapters. The chapters on chemistry, geoscience, energy and the environment, and engineering are relevant to the petroleum sciences. Each chapter is further divided according to type of source. A listing of online computer databases related to science and engineering literature is included as well as a selected bibliography for the information science student. The index provides access to the text by author, title, and subject. More than twelve hundred reference sources are discussed.

2 Indexing and Abstracting Services

One person could never hope to read or even scan the thousands of new petroleum-related journal articles, government documents, conference proceedings, books, and dissertations that are published each year. It is this explosion of literature that makes indexes and abstracts so important. Petroleum-related indexes and abstracts can be used to maintain current awareness, to locate a specific reference, or to locate materials in a general subject area.

The primary petroleum indexing and abstracting services are *Petroleum Abstracts* (entry 21) and *API Abstracts/Literature* (entry 10). These two services cover most petroleum journals, conference proceedings, dissertations, books, and patents. *Petroleum Abstracts* concentrates on petroleum exploration and production, while *API Abstracts* emphasizes refining and petrochemicals.

This chapter covers petroleum abstracts and indexes. Materials from peripheral sciences, such as *Chemical Abstracts* (entry 13), *Corrosion Abstracts* (entry 14), and *The Engineering Index Annual* (entry 15) have also been included in an attempt to show the interdisciplinary nature of the petroleum sciences. Other sources, not annotated here, but of potential use to the petroleum researcher or information specialist, are *Physics Abstracts*,[1] *Science Citation Index*,[2] and *Energy Research Abstracts*.[3]

Many indexes, abstracts, and bibliographies have a computerized counterpart. A full discussion of these online sources is found in chapter 8.

---

[1]*Physics Abstracts* (Science Abstracts, Section A) (Piscataway, N.J.: INSPEC, IEEE, 1898- . bimonthly. ISSN 0036-8091).

[2]*Science Citation Index* (Philadelphia: Institute for Scientific Information, 1955- . 6 issues per year. ISSN 0035-827X).

[3]*Energy Research Abstracts* (Oak Ridge, Tenn.: U.S. Department of Energy, Technical Information Center, 1976- . semimonthly. ISSN 0160-3604).

9. **API Abstracts/Catalysts and Catalysis.** New York: American Petroleum Institute, 1984- . weekly. $95.00/yr.

This publication includes abstracts on the "manufacture and use of catalysts for petrochemical processes; petroleum and other fossil fuel processes; and processes for the control of pollution." Coverage is worldwide. This is an important current awareness tool, especially for those involved in catalyst research and development.

10. **API Abstracts/Literature.** New York: American Petroleum Institute, 1978- . weekly. $18,300.00/yr.

These abstracts cover international petroleum-related articles in the fields of chemistry, physics, engineering, and alternate energy sources. The abstracts are divided into four major subject categories: *Petroleum Refining and Petrochemicals Literature Abstracts* (weekly); *Abstracts of Health and Environment Literature* (weekly); *Abstracts of Transportation and Storage Literature* (monthly); *Abstracts of Petroleum Substitutes Literature* (monthly). Access to these abstracts is a "must" for any petroleum library or information center. Also available as a computerized database called APILIT (entry 236).

11. **API Abstracts/Oilfield Chemicals.** New York: American Petroleum Institute, 1981- . monthly. $640.00/yr.

This service abstracts journal articles, conference papers, and patents related to chemicals used in oilfield applications. The scope is international. Each issue is divided into sections covering drilling fluids, completion and stimulation fluids, enhanced recovery and economics, production chemicals, and statistics. This current awareness tool is valuable to researchers, students, and librarians alike.

12. **API Abstracts/Patents.** New York: American Petroleum Institute, 1978- . weekly. $7,500.00/yr.

"This is a weekly service that provides abstracts of patents significant to the petroleum refining and petrochemical industries with extensive coverage in the chemicals and polymers areas." The abstracts are divided into seven groups: *Petroleum Processes; Primary and Specialty Products; Petroleum Substitutes; Chemical Products; Polymers; Conservation – Transportation – Engineering – Storage;* and *Agriculturals.* Also available as a computerized database called APIPAT (entry 237).

13. **Chemical Abstracts.** Columbus, Ohio: Chemical Abstracts Service, 1907- . weekly. $9,200.00/yr. (base subscription price). LC 09-4698. ISSN 0009-2258.

Chemical Abstracts (*CA*) is considered to be the most important abstracting service for the world's chemical literature. Currently, *CA* is processing about eight thousand documents weekly from twelve thousand publications in fifty languages. The abstracts are available in a printed format, on microfilm, or on microfiche. A cumulative printed index is issued every ten years. The same bibliographic information is also available online from 1967 to the present from a variety of vendors. *CA* is particularly useful when searching for petroleum-related articles and patents in the fields of enhanced oil recovery, petroleum processing and refining, drilling fluids, and petroleum products.

14. **Corrosion Abstracts.** Houston, Tex.: National Association of Corrosion Engineers, 1962- . bimonthly. $200.00/yr. (individuals); $250.00/yr. (libraries). LC 64-39093. ISSN 0010-9339.

2 – INDEXING AND ABSTRACTING SERVICES / 7

Many abstracting services permit the National Association of Corrosion Engineers (NACE) to reprint their abstracts in this publication. A few original NACE abstracts are also included. *Corrosion Abstracts* covers the worldwide corrosion control literature. It is an important reference work for the petroleum industry because it includes abstracts on corrosion associated with production, pipelines, and refining. Each issue of these abstracts contains an alphabetical subject index. A combined annual subject index is published in the last issue of the year.

15. **The Engineering Index Annual.** New York: Engineering Information, Inc. 1906- . annual. $745.00/yr. LC 73-8575. ISSN 0360-8557.

Provides worldwide coverage of the technical literature in all engineering disciplines. This index includes bibliographic citations and abstracts for journal articles, conference proceedings, books, technical reports, and other engineering materials. *The Engineering Index* is particularly useful for references in the areas of petrochemicals, petroleum analysis, petroleum reservoir engineering, pipelines, petroleum products, and petroleum refining. *The Engineering Index Annual* cumulates the monthly issues. Compendex is the online database equivalent of *The Engineering Index Monthly and Author Index* (ISSN 0162-3036). The database includes records from 1970 to the present and is available through several major vendors.

16. **Fuel and Energy Abstracts.** Vol. 19- . Surrey, England: Butterworth Scientific Ltd., 1978- . bimonthly. $260.00/yr. LC 79-641513. ISSN 0140-6701.

These abstracts summarize the worldwide literature on fuel and energy. Entries are listed in twenty-seven broad subject classes and are further arranged alphabetically by title. Most titles are followed by a complete abstract. Reference to the original source of the abstract is included with each entry. This publication was formerly called *Fuel Abstracts and Current Titles* (ISSN 0016-2388. Vol. 1-18, 1960-1977).

17. **Gas Abstracts.** Vol. 1- . Chicago: Institute of Gas Technology, 1945- . monthly, with annual index. $125.00/yr. ISSN 0016-4844.

These abstracts include "selected articles, papers, patents and other materials of interest to the gas and energy industries...." Order information is provided for articles which are not available from the Institute of Gas Technology (IGT) document delivery service. Publications from two hundred periodicals and over fifty associations are surveyed for inclusion in this abstracting service.

18. **General Index to Publications of the Society of Petroleum Engineers of AIME.** Dallas, Tex.: Society of Petroleum Engineers of AIME, 1954- . irregular. contact publisher for price.

The Society of Petroleum Engineers (SPE) has published four volumes of the *General Index* since 1954. These publications provide indexes by author and subject for papers published by SPE since 1921. Volumes 3 and 4 of the index include papers appearing in *Society of Petroleum Engineers Journal* and the *Journal of Petroleum Technology*. Each volume is listed below with its respective ISBN and price. Copies are available from SPE.

Vol. 1, 1921-1952. ISBN 0-89520-212-3. $8.00 (members); $12.00 (nonmembers).
Vol. 2, 1953-1966. ISBN 0-89520-211-5. $6.00 (members); $7.75 (nonmembers).
Vol. 3, 1967-1974. ISBN 0-89520-210-7. $7.00 (members); $10.00 (nonmembers).
Vol. 4, 1975-1980. ISBN 0-89520-312-X. $15.00 (members); $22.50 (nonmembers).
Vol. 5, 1981-1985. Expected publication date 1986.

8 / 2—INDEXING AND ABSTRACTING SERVICES

19. **International Petroleum Abstracts.** Vol. 1- . Sussex, England: John Wiley & Sons Ltd., 1972- . quarterly. $190.00/yr. LC 73-646769. ISSN 0309-4944.

This abstracting service surveys a highly selected list of petroleum-related publications. Each volume contains a list of the journals abstracted, an annual subject index, and an author index. Patent literature is not included. Some of the major topics covered are oilfield exploration and exploitation, refining and related processes, petroleum products, and economics. These abstracts are most useful when an overview of the petroleum literature is needed. They should not, however, be considered comprehensive.

20. **Offshore Abstracts.** Vol. 1- . London: Offshore Information Literature, 1974- . bimonthly. £130.00/yr. (outside Great Britain). LC 85-14416. ISSN 0305-0513.

This service "provides extensive review of information for the Offshore Industry, from a world-wide coverage of scientific and technical journals, conference papers, research reports, trade literature, standards and patents." The abstracts are arranged according to subject categories, including buoys and mooring systems, cathodic protection, corrosion, drilling, pipes and pipelines, platforms and rigs, production, and structural engineering.

21. **Petroleum Abstracts.** Vol. 1- . Tulsa, Okla.: University of Tulsa, 1961- . weekly. contact publisher for price. ISSN 0031-6423.

Petroleum Abstracts is one of the most important current awareness sources available to the petroleum exploration and production industry. Materials are abstracted from more than five hundred U.S. and foreign periodicals, patent journals, meeting papers, and government documents. Each issue arranges the abstracts by broad subject categories, including geology, geochemistry, geophysics, drilling, well logging, well completion and servicing, production of oil and gas, reservoir engineering and recovery methods, pipelining, shipping and storage, ecology and pollution, alternate fuels and energy sources, and mineral commodities. The same information found in the printed version is available online in the TULSA database (entry 248). Single copies of most articles and patents abstracted are available from Petroleum Abstracts Duplicating Services, McFarlin Library, 600 South College Avenue, Tulsa, OK 74104.

22. **Petroleum/Energy Business News Index.** New York: American Petroleum Institute, 1975- . monthly. $610.00/yr.; $310.00/yr. (for basic supporters of the *API Technical Index*). LC 75-645032. ISSN 0098-7743.

Sixteen petroleum industry journals are indexed cover-to-cover in this monthly publication. Approximately thirty-three hundred articles and news announcements are included in each month's issue. Coverage includes plans and activities of government agencies and corporations, environmental matters, economic data, and news about people in the petroleum industry. This same information is available online in the P/E News database (entry 247).

23. **Society of Petroleum Engineers Technical Papers.** Ann Arbor, Mich.: University Microfilms International, 1981. 2v. Volume 1 **Author and Title Indexes to the Microfiche Collection.** Volume 2 **Subject and Fiche Number Indexes.** $300.00/set (nonmembers); $270.00/set (members). LC 80-28727. ISBN 0-8357-0217-0 (set).

This two-volume set provides access to over eight thousand papers presented at meetings sponsored by the Society of Petroleum Engineers (SPE) from 1957 to 1979. These papers are available in microfiche form, and this index is designed to provide author, title, subject, and fiche number access to the microfiche. Papers are listed in sequential order by fiche number. Each year, as new papers are added to the fiche collection, a single-volume supplement updates the two-volume index. Annual supplements are available to SPE members at a cost of $31.50 and to non-SPE members for $35.00. Although this index was designed for use with the microfiche collection of SPE papers, anyone interested in identifying specific papers presented at SPE meetings will find this a useful reference tool. Copies of individual SPE papers can be purchased from the society.

3 Dictionaries

The language of the petroleum industry is unique. Words like *tour, fishing, pig,* and *logs* have entirely different meanings when used in the context of the oilfield. The sources listed in this chapter identify and explain this unique terminology. Some dictionaries focus on a specific aspect of petroleum technology, while others attempt to give an overview of the entire industry. In addition to definitions, petroleum dictionaries may include conversion factors, lists of abbreviations, SI units, style manuals, maps, and symbols.

Dictionaries have been arranged in three main categories: general works, specialized, and multilingual. Entries are listed alphabetically by title within each category. *The Illustrated Petroleum Reference Dictionary* (entry 27) and *The Petroleum Dictionary* (entry 28) are excellent examples of general dictionaries. Both are broad in scope and well illustrated, and appeal to the nontechnical as well as the technical user. Specialized sources include subject-specific lexicons, dictionaries of abbreviations, and aids for translation. Subject-specific titles such as *Encyclopedic Dictionary of Exploration Geophysics* (entry 35), *Glossary of Geology* (entry 36), and *An A-Z of Offshore Oil & Gas* (entry 32) provide in-depth coverage of a single aspect of oil and gas technology. *D & D Standard Oil Abbreviator* (entry 33) is considered the classic source for deciphering petroleum-related abbreviations. Because of its popularity and usefulness, it has been reprinted in other petroleum reference sources.

The international nature of the petroleum industry has created a need for tools that translate oil and gas terminology into many languages. These multilingual aids provide foreign-language equivalents but rarely include definitions. *Elsevier's Oil and Gas Field Dictionary in Six Languages* (entry 41) is the most comprehensive multilingual dictionary. It includes translations from American English into six foreign languages. Most of the other dictionaries in this chapter that include foreign-language equivalents are bilingual.

General Works

24. **The Dictionary of Energy Technology.** By Alan Gilpin and Alan Williams. London: Butterworth Scientific Ltd., 1982. 392p. illus. $49.95. LC 80-42357. ISBN 0-408-01108-4.

Many basic oil and gas terms are included in this general dictionary of energy terms. The dictionary covers a wide range of topics. Specific oilfield terminology is usually not included. Short lists of common and scientific abbreviations, conversion factors, and a brief energy bibliography are also included.

25. **A Dictionary of Petroleum Terms.** 3rd ed. Austin, Tex.: Petroleum Extension Service at the University of Texas at Austin, 1983. 177p. $16.00. LC 83-61652. ISBN 0-88698-000-3.

This dictionary of oilfield terminology is a useful tool for anyone attempting to learn basic oilfield terms or for those who need to verify spellings or definitions of specific words. Italicized words within a definition indicate that that word is also defined in the dictionary. Petroleum industry abbreviations and a list of metric equivalents are included.

26. **Handbook of Oil Industry Terms and Phrases.** 4th ed. By R. D. Langenkamp. Tulsa, Okla.: PennWell Publishing Co., 1984. 347p. $29.95. LC 84-5837. ISBN 0-87814-258-4.

Oil industry terminology is dynamic, so an up-to-date list of vocabulary is very important. All of the terms in this handbook have been defined using nontechnical language. Words to be defined are typed in uppercase letters and are followed by two-or three-sentence definitions. Some previous knowledge of petroleum, geological, geophysical, etc., terminology makes the definitions easier to understand. A good tool for libraries or offices.

27. **The Illustrated Petroleum Reference Dictionary.** 3rd ed. Edited by Robert D. Langenkamp. Tulsa, Okla.: PennWell Publishing Co., 1985. 696p. illus. $55.95. LC 78-71326. ISBN 0-87814-272-X.

This dictionary includes definitions for over thirty-five hundred oil or petroleum-related terms. Many definitions are accompanied by excellent photographs or drawings. Also included are the Desk and Derrick Clubs' list of standard oil abbreviations and Steven Gerolde's *Universal Conversion Factors* (Tulsa, Okla.: PennWell Publishing Co., 1971. 276p. $19.95. LC 71-164900. ISBN 0-87814-005-0). Anyone who needs to understand basic petroleum terminology or jargon will find this an excellent reference tool.

28. **The Petroleum Dictionary.** By David F. Tver and Richard W. Berry. New York: Van Nostrand Reinhold, 1980. 374p. illus. $24.50. LC 79-19346. ISBN 0-442-24046-5.

This is a comprehensive petroleum dictionary covering "geology, geophysics, seismology, drilling, gas processing, production, a detailed analysis of various refining operations and processes, offshore technology, and description of various materials and supporting techniques used in drilling and production of oil and gas." It includes several good drawings and diagrams. This dictionary should be useful to anyone associated with the oil and gas industry, including students, technicians, researchers, and information specialists.

29. **Petroleum Industry Glossary.** Edited by Susan Reeves Palmer. Oklahoma City, Okla.: IED, 1982. 272p. $32.00. ISBN 0-89419-289-2.

This work is a general, nontechnical compilation of petroleum industry terminology. Words or phrases that are underlined in a definition refer to another definition for additional information. For example, the definition of *geophones* refers the reader to *hydrogeophones*. Appendices include abbreviations, well classification guidelines, map symbols, rock symbols, the geologic time scale, and geophysical exploration and well logging symbols. A bibliography is also included.

30. **Phillips 66 Glossary: Selected Words and Phrases Used in the Energy and Petrochemical Industries.** Bartlesville, Okla.: Phillips Petroleum Co., n.d. 157p. free.

No attempt was made to be comprehensive in this brief dictionary of common petroleum words and phrases. The main purpose of the glossary is to provide a quick reference tool for those who are unfamiliar with petroleum-related language. The publication is available at no charge from Public Relations, Phillips Petroleum Company, 16-A2 Phillips Building, Bartlesville, OK 74004.

31. **The World Energy Book; An A-Z, Atlas and Statistical Source Book.** By David Crabbe and Richard McBride. New York: Nichols Publishing Co., 1978. 259p. illus. $32.50. LC 78-50805. ISBN 0-89397-032-8.

This multipurpose reference book is divided into three sections: "A-Z," the "Energy Resources Atlas," and "Statistical Appendices." The A-Z section is an illustrated dictionary of approximately one thousand energy terms. Many terms are cross-referenced to related terms or to information in the atlas section. The "Energy Resources Atlas" includes maps of oil and gas field locations; coal, heavy oil, and uranium deposits; ocean energy sources; geothermal areas; and solar or hydroelectric energy potential. The "Statistical Appendices" section includes very basic, ten-year (usually 1966-1976) statistics on energy production, reserves, consumption, and capacity. Conversion tables, calorific equivalents and values, the geologic time scale, and the classification of crude petroleum are presented in tabular form.

Specialized

32. **An A-Z of Offshore Oil & Gas: An Illustrated International Glossary and Reference Guide to the Offshore Oil & Gas Industries and Their Technology.** 2nd ed. By Harry Whitehead. Houston, Tex.: Gulf Publishing Co., 1983. 438p. illus. $44.50. LC 82-84656. ISBN 0-87201-052-X.

The first part of this reference book contains definitions of words used in the offshore oil and gas industry. Illustrations are included for some definitions. Italicized words within the definitions indicate related terms or additional information found either in the glossary or in one of the appendices. The second part of the book is a set of thirty appendices. Information in this section is primarily presented by means of tables, maps, or diagrams. A wide range of topics is covered, including offshore areas of exploration and production, types of equipment or vessels used in the offshore industry, abbreviations commonly encountered, refining information, and petrochemical data.

33. **D & D Standard Oil Abbreviator.** 2nd ed. Tulsa, Okla.: PennWell Publishing Co., 1973. 230p. $14.95. LC 72-96172. ISBN 0-87814-017-4.

This glossary of abbreviations was compiled by the Association of Desk and Derrick Clubs. Originally written for secretaries, draftsmen, and engineers, it is useful to anyone who has to decipher petroleum-related abbreviations. It is divided into six sections: (1) abbreviations listed alphabetically followed by definitions; (2) definitions

3 – DICTIONARIES / 13

listed alphabetically followed by abbreviations; (3) abbreviations for logging tools and services; (4) abbreviations for U.S. and Canadian petroleum companies and associations; (5) abbreviations for companies and associations outside the United States and Canada; and (6) miscellaneous map symbols, mathematical symbols, and conversions.

34. **Dictionary of Geological Terms.** 3rd ed. By Robert L. Bates and Julia A. Jackson. Garden City, N.Y.: Anchor Press/Doubleday, 1984. 571p. $19.95. LC 82-45315. ISBN 0-385-18100-0.

This dictionary is intended for use by nonspecialists, earth sciences students, and teachers rather than by professional earth scientists. Terms in the dictionary focus on an entry-level understanding of the geological sciences. An effort is made to avoid jargon or highly technical terms. Those seeking a more inclusive, technically oriented dictionary are referred to the *Glossary of Geology*, also by Bates and Jackson (see entry 36). Terms and phrases are arranged alphabetically, letter-by-letter, and include preferred pronunciations as well as definitions. Italicized words within a definition indicate words that are defined elsewhere in the dictionary.

35. **Encyclopedic Dictionary of Exploration Geophysics.** 2nd ed. By Robert E. Sheriff. Tulsa, Okla.: Society of Exploration Geophysicists, 1984. 323p. $20.00. LC 84-051971. ISBN 0-931830-31-8.

This reference tool is intended to be useful to the practicing geophysicist. Entries are arranged alphabetically and, in addition to the definition, often include a brief discussion of the topic, citations to related literature, alternate meanings, and synonyms. Where appropriate, cross-references are made within a definition to terms defined elsewhere in the dictionary. Fourteen appendices provide further information, including symbols used in geophysics, a geologic time scale, conversion units, the Greek alphabet, and instructions to authors from the editors of the journal *Geophysics*.

36. **Glossary of Geology.** 2nd ed. By Robert L. Bates and Julia A. Jackson. Falls Church, Va.: American Geological Institute, 1980. 749p. $60.00. LC 79-57360. ISBN 0-913312-15-0.

The *Glossary of Geology* has been carefully compiled and reviewed by nearly 150 specialists representing thirty-two fields in the geosciences. It is intended to be an authoritative reference source for the professional earth scientist. Arranged alphabetically, letter-by-letter, many definitions include citations to references for further reading. The bibliography for these citations is included at the end of the book. Italicized words indicate terms which have been defined elsewhere in the dictionary. Geoscientists and information specialists working in geology or petroleum-related fields will find this glossary an essential part of their office reference collections.

37. **Oil and Gas Terms: Annotated Manual of Legal Engineering Tax Words and Phrases.** 6th ed. By Howard R. Williams and Charles J. Meyers. New York: Matthew Bender, 1984. 985p. $60.00. LC 85-138996.

Terms in this manual are arranged alphabetically, letter-by-letter, and cover all aspects of activity in the petroleum industry from exploration to production to transportation. The dictionary is intended primarily for lawyers, landmen, students, and accountants needing clarification of petroleum-related terminology. Many entries contain citations to statutes, cases, books, or law review articles that may further interpret the definition or the term.

14 / 3 — DICTIONARIES

38. **Standard Definitions for Petroleum Statistics (Technical Report No. 1).** Washington, D.C.: American Petroleum Institute, 1981. 46p. index. $5.00.

This work defines terms used in statistical reporting of oil and gas exploration and production activities. The terms are divided into three sections: part 1 covers petroleum reserves and production, part 2 includes terms related to wells and drilling, and part 3 discusses products and refining. Four appendices contain maps and flow charts related to petroleum industry statistics.

Multilingual

39. **Dictionary of Geosciences.** Edited by Adolf Watznauer. New York: Elsevier Science Publishing Co., 1982. 2v. $57.50/vol. LC 81-22071. ISBN 0-444-99701-6 (vol. 1, German-English); 0-444-99702-4 (vol. 2, English-German).

This bilingual dictionary, a translation of *Wörterbuch Geowissenschaften*, is intended to help those needing to know the English and German equivalents for geoscience terminology. Over thirty-eight thousand terms are presented in alphabetical arrangement. Two tables, a stratigraphic table and a seismic intensity scale, are also included.

40. **Dictionary of Petroleum Technology (Dictionnaire technique du pétrole).** By Magdeline Moureau and Gerald Brace. Paris: Éditions Technip, 1979. 946p. $95.00. LC 79-188223. ISBN 2-7108-0361-5.

Approximately fifty thousand terms and expressions are listed in this bilingual petroleum dictionary. "Definitions are given for basic concepts or terms of difficult access in the standard literature, or for ambiguous expressions." Translations are provided for all terms even if the term is not fully defined. The dictionary is divided into two sections, English-French and French-English. Appendices contain conversion tables, charts, and symbols.

41. **Elsevier's Oil and Gas Field Dictionary in Six Languages.** By L. Y. Chaballe, L. Masuy, J. P. Vandenberghe, and S. Salem. Amsterdam: Elsevier Scientific Publishing Co.; distr., New York: Elsevier Science Publishing Co., 1980. 672p. indexes. $149.00. LC 80-11791. ISBN 0-444-41833-4.

This dictionary includes more than forty-eight hundred terms and emphasizes terminology in the areas of geology, petroleum engineering, oilfield production, offshore, hydrocarbon exploration, and oil or gas transportation. The two-part dictionary consists of a basic table and indexes. Entries in the basic table are arranged alphabetically in English and are followed by the equivalent word in French, Spanish, Italian, Dutch, and German. Indexes for each of these languages follow the basic table. Entries in the indexes have a numeric code, which refers to the appropriate English word in the basic table. An Arabic supplement is included at the end of the dictionary.

42. **English-Spanish and Spanish-English Glossary of the Petroleum Industry. (Glosario de la industria petrolera).** 2nd ed. Tulsa, Okla.: PennWell Publishing Co., 1982. 378p. $29.95. LC 82-7622. ISBN 0-87814-194-4.

This glossary contains over fourteen thousand technical terms related to the petroleum industry. Terms are translated from English to Spanish in the first half, and from Spanish to English in the second half. This tool is primarily a translation aid — terms are not defined. It covers equipment, processes, and techniques used in all phases of the oil and gas industry.

43. **Onshore/Offshore Oil & Gas Multilingual Glossary.** London: Graham & Trotman Ltd., 1979. 490p. indexes. $44.00. LC 81-107553. ISBN 0-86010-184-3.

This glossary is intended for translators, interpreters, and researchers. It has indexes in English, Danish, German, French, Italian, and Dutch. Unlike *Elsevier's Oil and Gas Field Dictionary in Six Languages* (entry 41), which provides word-for-word translations, this dictionary attempts to indicate appropriate word use in context. Each indexed word refers to an entry number where the word is used in context in a sentence. The sentence is repeated in each of the six languages. A bibliography of sources for the translated sentences is provided at the end of the glossary.

44. **Petroleum Drilling Equipment, Terms & Phrases English-Spanish, Spanish-English.** By Arthur E. Thomann. New York: Marlin Publications International, 1980. 423p. $50.00. LC 80-82143. ISBN 0-930624-02-5.

Because of its focus on oil industry equipment, this reference source contains translations for many terms which are not defined in other bilingual petroleum dictionaries. Each term is followed by the part of speech and the field or specialty to which the term relates. The corresponding term in the second language is then identified. All terms are translated from English to Spanish and from Spanish to English. Multiple indentations and dashes make this dictionary a little difficult to use. Previously published under the title *Glossary of Petroleum Industry Equipment Terms & Phrases.*

45. **Les termes pétroliers dictionnaire anglais-français.** By Michel Arnould and Fabio Zubini. London: Graham & Trotman Ltd., 1981. 288p. $30.00. LC 82-167169. ISBN 0-86010-374-9.

This English-French dictionary includes more than eighty-seven hundred words or phrases related to the petroleum industry. Words or phrases are first given in English and then in French.

4 Encyclopedias and Yearbooks

There are few traditional encyclopedias among petroleum reference sources. The best-known petroleum "encyclopedia," the *International Petroleum Encyclopedia* (entry 50), is actually a yearbook. It is published annually, arranged by geographic area, and focuses on data and statistics.

Petroleum encyclopedias are usually subject-specific and provide in-depth information. For example, there are a well logging encyclopedia (entry 47), a natural gas encyclopedia (entry 48), and a land leasing encyclopedia (entry 51). Other encyclopedias that are useful to the petroleum industry include sources such as the *Encyclopedia of Earth Sciences Series* (entry 46), *Encyclopedia of Chemical Technology*,[1] and the *Encyclopedia of Science & Technology*.[2] These sources provide substantive articles written by experts, and include bibliographies for more in-depth study.

Petroleum yearbooks, like those in most other disciplines, present general industry overviews, statistics, and chronologies of significant events for the year. In addition, many of these yearbooks contain updated directory information. This is especially true of the *European Petroleum Year Book* (entry 53), the *National Petroleum News Fact Book* (entry 52), and the *Arab Oil & Gas Directory* (entry 61).

The annotations in this chapter are arranged geographically. Subdivisions include general works, Europe, Great Britain, Japan, Middle East, and the United States.

---

[1] *Encyclopedia of Chemical Technology* (New York: John Wiley & Sons, 1984). LC 77-15820. ISBN 0-471-80104-6.

[2] *Encyclopedia of Science & Technology* (New York: McGraw-Hill Book Co., 1982). LC 81-20920. ISBN 0-07-079280-1.

General Works

46. **Encyclopedia of Earth Sciences Series.** Edited by Rhodes W. Fairbridge. New York: Van Nostrand Reinhold, 1966- . irregular. price varies.

This series attempts to present a comprehensive overview of the earth sciences in a multivolume, encyclopedic format. Articles in each volume are individually written by subject experts and include reference lists for additional reading. Volume 1, *The Encyclopedia of Oceanography*, was published in 1966 and the latest volume, on beaches and coastal environments, was published in 1982. Because of this sixteen-year time span, some of the material in earlier volumes is out-of-date. Much of the historical information, however, is still quite useful. Each volume in this actively growing series is listed below with its own ISBN.

Volume 1, *Encyclopedia of Oceanography*, 1966. ISBN 0-442-15070-9.

Volume 2, *Encyclopedia of Atmospheric Sciences and Astrogeology*, 1967. ISBN 0-442-15071-7.

Volume 3, *Encyclopedia of Geomorphology*, 1975. ISBN 0-12-7864-59-8.

Volume 4a, *Encyclopedia of Geochemistry and Environmental Sciences*, 1972.

Volume 4b, *Encyclopedia of Mineralogy*, 1981. ISBN 0-87933-184-4.

Volume 6, *Encyclopedia of Sedimentology*, 1978. ISBN 0-87933-152-6.

Volume 7, *Encyclopedia of Paleontology*, 1979. ISBN 0-87933-185-2.

Volume 8, *Encyclopedia of World Regional Geology, Part 1*, 1975. ISBN 0-470-25145-X.

Volume 12, *Encyclopedia of Soil Science, Part 1*, 1979. ISBN 0-87933-176-3.

Volume 15, *Encyclopedia of Beaches and Coastal Environments*, 1982. ISBN 0-87933-422-3.

47. **Encyclopedia of Well Logging.** By Robert Desbrandes. Houston, Tex.: Gulf Publishing Co., 1985. 584p. index. illus. $89.00. LC 85-70854. ISBN 0-87201-249-2.

This encyclopedia includes a wealth of information on all types of well logging tools, procedures, and techniques. Each chapter is well illustrated and most contain a list of references for further reading. Some of the logging techniques that are discussed are electric, dielectric, nuclear, acoustic, and wireline. Several chapters discuss logging in special environments. The appendices include a listing of conversion factors, a brief bibliography on well logging, a listing of schools and universities offering graduate courses in well logging, and a five-language glossary of well logging terms. This is a good reference source for students, trainees, engineers, and information specialists.

48. **Encyclopedie des gaz; Gas Encyclopaedia.** By L'Air Liquide, Division Scientifique (English translation by Nissim Marshall). Amsterdam: Elsevier Scientific Publishing Co., 1976. 1150p. index. $191.50. ISBN 0-444-41492-4.

Presented in French and in English, this encyclopedia includes data on 138 gases. General information on gas properties, flammability of gases, handling, and metrology of gases is presented in the first five chapters. Specific data on the gases follow in individual "monographs." The following information is outlined in each "monograph": physical properties, flammability, biological properties, safe handling and storage procedures, leak detection, materials of construction, uses, and a bibliography.

49. **International Oil and Gas Development.** Vol. 1- . Austin, Tex.: International Oil Scouts Association, 1931- . annual. index. price varies. LC 32-22187.

18 / 4—ENCYCLOPEDIAS AND YEARBOOKS

This yearbook reviews the exploration and production of oil and gas fields in the United States and Canada. Exploration data and brief production overviews are also included for a few countries outside of North America. Exploration reviews and production reviews are published in separate volumes. Statistics tend to lag three to four years behind the publication date of the yearbook. Canadian statistics are listed by province and field, U.S. figures by state, county, and field. Exploration entries include geological, geophysical, land leasing, wildcat, and development information. Production reviews include field discovery date, producing formation, API gravity, depth, number of wells, and production figures. For historical data, this yearbook is an essential tool for librarians or information centers in petroleum-related companies.

50. **International Petroleum Encyclopedia.** Vol. 1- . Tulsa, Okla.: PennWell Publishing Co., 1968- . annual. $75.00/yr. LC 77-76966. ISBN 0-87814-282-7 (1985 ed.). ISSN 0148-0375.

Presented in yearbook format, this encyclopedia contains current information on petroleum-related activities in countries throughout the world. Maps are included which show locations of oilfields, gas fields, pipelines, refineries, and tanker terminals. Production and refining statistics are given. Graphs, tables, and photographs enhance review articles and discussions covering major topics of interest to the petroleum researcher.

The 1985 volume includes sections titled "Heavy Oil," "Production by Fields," "Worldwide Well Completions," and "Worldwide Oil and Gas Production at a Glance." An international directory is also included in this issue, listing name, address, telephone number, telex number, and cable number for petroleum agencies and associations.

This is a valuable tool for the information specialist, researcher, or student in identifying trends in technology and petroleum development by country. A brief table of contents and an index are provided.

51. **Landman's Encyclopedia.** 2nd ed. By R. L. Hankinson and R. L. Hankinson, Jr. Houston, Tex.: Gulf Publishing Co., 1981. 494p. index. $75.00. LC 81-6422. ISBN 0-87201-420-1.

Intended primarily for use by the practicing petroleum landman, this reference tool provides copies of forms and instruments used in the leasing of property for oil and gas drilling and exploration activities. In addition to the many sample forms, contracts, and agreements, there are also explanations of procedures to be followed in specific circumstances. A topical index is included.

52. **National Petroleum News Fact Book.** Des Plaines, Ill.: Hunter Publishing Co., 1954- . annual. index. $35.00/yr. LC 78-132. ISSN 0149-5267.

This is a special annual issue of *National Petroleum News* that provides a comprehensive look at the petroleum industry for the past year. The *Fact Book* includes financial data, market surveys, statistics on supply, demand, and reserves, and directory information on marketing/management companies and petroleum-related associations. The *Fact Book* is exactly what its name implies—a compilation of yearly facts and statistics concerning all aspects of the petroleum industry.

Europe

53. **Annuaire européen du pétrole. (European Petroleum Year Book).** Hamburg, West Germany: Otto Vieth Verlag, 1963- . annual. indexes. DM 182,00/yr. LC 80-2643. ISSN 0342-6947.

The seventeenth edition of *Annuaire européen du pétrole* presents a combination of yearbook and directory information. Four individually numbered sections are titled "Editorial Topics," "Oil and Gas in Europe," "Companies' Directory," and "Suppliers' Directory & Buyers' Guide." Section 1, "Editorial Topics," includes two signed articles on themes of current interest. Articles are presented in English, French, and German. Section 2, "Oil and Gas in Europe," includes information on reserves, production, operator, and year of discovery for individual fields. Entries are arranged by country and then alphabetically by field name. Excellent maps showing field locations, pipelines, and refineries are found in this section. European oil and gas statistics complete section 2. Section 3, "Companies' Directory," gives addresses, telephone numbers, telex numbers, and key personnel for exploration, production, refining, trading, marketing, distribution, holding, and pipeline operation companies in Europe. The fourth section is divided into the "Suppliers' Directory" and the "Buyers' Guide." The "Suppliers' Directory" is arranged first geographically and then by broad subject. The "Buyers' Guide" is arranged by product or service name and includes an English, French, and German index.

54. **European Offshore Oil & Gas Yearbook and Directory.** London: Thomas Telford Ltd., 1985- . annual. £40.00/yr. ISBN 0-7277-0229-7 (1985 ed.).

This work updates and expands the former *UK Offshore Oil and Gas Directory*. An excellent source for information on all European offshore efforts, it includes detailed offshore reviews for Denmark, France, Greece, Ireland, Italy, the Netherlands, Norway, Portugal, Spain, Sweden, the United Kingdom, and West Germany. Each review includes a status report, license terms and conditions, tax information, field activity, lists of organizations, and block maps. This publication also includes a brief discussion of the geology of northwest Europe, a glossary of offshore terminology, a selected guide to offshore publications, and a suppliers' directory.

55. **Fact Sheet: The Norwegian Continental Shelf.** Oslo, Norway: The Royal Ministry of Petroleum and Energy, 1984(?)- . annual. illus. contact publisher for price. ISSN 0800-7683.

This publication provides up-to-date general information on petroleum activity on the Norwegian continental shelf. Data presented in the 1985 edition include exploration and production activities, development projects, fields under consideration, offshore licensing activities, petroleum revenues, and information on the market situation for petroleum products. A list of operator addresses is included at the end of the yearbook. The excellent overview information and data on current Norwegian petroleum activity make this publication valuable to the student, practicing engineer, and information specialist. Copies are available upon request from The Royal Ministry of Petroleum and Energy, P.O. Box 8148 Dep., Oslo, Norway.

56. **Guide Offshore: French Offshore Yearbook.** Paris: Éditions Olivier Lesourd, 1981- . annual. Fr 250/yr. ISSN 0245-9310.

This bilingual yearbook is designed to reflect technological advances in the French offshore industry. The yearbook is divided into two major sections. Section 1 contains eighteen articles on deep-sea exploration. Each article is presented first in French, then in English. Section 2 provides directory information for governmental departments and research groups, oil contractors, professional associations, offshore-related insurance companies, petroleum-oriented studies, offshore contractors, equipment, and services. An alphabetical list of companies, with addresses and telephone numbers, is also included.

Great Britain

57. **Development of the Oil and Gas Resources of the United Kingdom.** London: Her Majesty's Stationery Office, 1985- . irregular. £9.50. ISBN 0-11-411565-6 (1985 ed.).

Information is provided on the economic, industrial, and environmental aspects of oil and gas activities in Great Britain. The 1985 edition of this publication also discusses exploration, reserves, development, and production efforts during 1984 and the early months of 1985. Further location data are contained on a definitive map of the United Kingdom continental shelf.

58. **Offshore Oil & Gas Yearbook: UK & Continental Europe.** Croydon, England: Benn Technical Books, 1979- . annual. illus. index. price varies. LC 79-642400. ISBN 0-85459-104-4 (1983-1984 ed.). ISSN 0141-5751.

This yearbook is divided into two distinct parts. "Part 1 contains principally country-by-country surveys of the offshore oil and gas exploration and production activities in Europe." Chapters in this section include an overview of the country, terms and conditions of licensing, national and official bodies, professional, trade, and employers' associations, licensees, field summaries, and block maps. "Part 2 is a directory of supplier companies and their products and services." An alphabetical list of companies at the end of part 2 provides readers with detailed information for companies listed under products and services categories found earlier in part 2. A key titled "Who Owns Whom" is also included.

Japan

59. **Japan Petroleum Industry Yearbook.** Tokyo: Japan International Consultants, Ltd., 1983- . annual. $180.00/yr. ISBN 4-930988-03-9 (1984 ed.).

Provides authoritative basic statistics for the Japanese petroleum industry. This yearbook is divided into four parts: "Government Petroleum Policy"; "Exploration and Development"; "Crude Oil and Petroleum Product Imports and Exports"; "Refining, Distribution and Marketing Activities"; and "Company Directory." Brief introductions precede many sections and provide overview information for procedures used in Japan's petroleum industry. Data presented include exploration and development statistics, crude imports, product imports and exports, refining and sales, petroleum product supply and demand, and product consumption.

Middle East

60. **Arab Business Yearbook.** London: Graham & Trotman Ltd., 1976- . annual. illus. $70.00/yr.; $56.00pa./yr. ISBN 0-86010-372-2; 0-86010-344-7(pa.). (1984 ed.). ISSN 0140-1874.

Provides interesting and informative data useful to people traveling and/or conducting business in the Arab countries. Part A includes information on population statistics, banking in the Arab world, advertising media, and the development of Arab oil economics. Part B is arranged alphabetically by country and presents valuable economic, financial, social, cultural, and general background information on twenty Arab countries. Entries for each country are concluded with a section on oil prospects, reserves, and economics.

61. **Arab Oil & Gas Directory.** Paris: The Arab Petroleum Research Center, 1974- . annual. $204.00/yr. LC 75-646597. ISBN 2-903282-04-8 (1985 ed.). ISSN 0304-8551.

This work surveys oil and gas industry activity in all Arab countries. Arrangement is alphabetical by country. The information is presented in yearbook format and encompasses the exploration, production, refining, and transportation activities within each country. Additional information is provided for other economic development projects underway in specific countries. Maps and tables support the text. An index lists the names and addresses of oil and service companies operating in the Middle East and North Africa. A valuable source for statistics and background information as well as discussions of current oil and gas development in this part of the world.

United States

62. **Resume: The Complete Annual Review of Oil and Gas Activity in the United States.** Denver, Colo.: Petroleum Information Corp., 1978- . annual. illus. $80.00/yr. ISSN 0270-7527.

Published in yearbook format, the *Resume* provides "narrative overviews of the year's activity, statistical data and comparisons, and detailed discovery well data ..." for the U.S. oil and gas industry. Narrative overviews cover exploration, drilling, production, expenditures, and corporate acquisitions as well as regional reviews and county statistical summaries. Excellent tables, graphs, and maps accompany narrative articles. Discovery well data are listed by region in an extensive appendix. This is a good source for current statistics.

63. **U.S. Energy Industry Yearbook.** New York: Whitney Communications Corp., 1977- . annual. index. $75.00/yr. ISBN 0-918216-10-9 (1985 ed.).

Includes up-to-date information on major integrated oil companies, drilling and exploration companies, independent petroleum companies, major petrochemical companies, and terminal companies. Each entry contains address, telephone number, key personnel, short financial statement, brief history, and current news. Photographs of the company's CEO are included with most entries. A section of petroleum-related federal agencies, trade associations, banks, and statistics is included at the end of the yearbook.

5 Handbooks, Manuals, and Basic Texts

Handbooks and manuals are intended to provide quick answers, and typically include data on a single subject. These reference books may include tables, charts, graphs, formulas, statistics, pertinent vocabulary, and brief subject overviews. Petroleum-related handbooks and manuals often provide specific information on production, consumption, reserves, procedures, material properties, or industry standards.

Two classic petroleum handbooks are *Petroleum Production Handbook* (entry 123) and *The Petroleum Handbook* (entry 73). These sources provide an excellent starting point for general petroleum facts and overview data. Examples of popular petroleum manuals such as *Air & Gas Drilling Manual* (entry 82) and the *IADC Drilling Manual* (entry 89) are also included in this chapter.

Another excellent type of tool, not considered "reference" literature in the pure sense, is the basic textbook or primer. Like handbooks and manuals, basic texts can provide useful charts, formulas, statistics, vocabulary, and, of course, subject overviews. Selected basic texts and primers are included in this chapter for those who need to familiarize themselves with the vocabulary and general workings of a specific petroleum subject area.

Entries for handbooks, manuals, and basic texts are arranged in the following categories: general works, business, drilling, exploration, legal, offshore, pipeline, production, refining, and style manuals.

5 – HANDBOOKS, MANUALS, AND BASIC TEXTS / 23

General Works

64. **Energy Handbook.** 2nd ed. By Robert L. Loftness. New York: Van Nostrand Reinhold, 1984. 763p. illus. index. $72.50. LC 83-21834. ISBN 0-442-25992-1.

Although most of this handbook deals with energy sources other than oil or natural gas, it also includes useful historical, statistical, and encyclopedic information about the petroleum industry. Chapter 2, "Fossil and Mineral Energy Resources," chapter 4, "Energy Consumption Trends," chapter 6, "Recovery of Fossil Fuels," and chapter 15, "Energy Costs," are probably the most useful for petroleum information specialists. These chapters include information on trends in crude oil prices, trends in U.S. fuel prices, tar sands recovery, oil shale recovery, exploration and production statistics, petroleum reserves, and much more. Graphs, tables, charts, and photographs are used extensively throughout. Other topics included in the handbook are renewable energy resources, nuclear energy, geothermal energy, solar energy, energy conversion, storage and transport, and environmental aspects of energy use. Energy conversion factors and a glossary are presented in two appendices.

65. **Fundamentals of Petroleum.** 2nd ed. Austin, Tex.: Petroleum Extension Service of the University of Texas at Austin, 1981. 1v. $10.00. LC 81-622847.

Intended for the layperson as well as the professional, this text provides an excellent overview of the petroleum industry. Chapter 1 begins with basic concepts pertaining to petroleum geology and reservoirs, and subsequent chapters discuss exploration, drilling, production, pipeline operations, refining, and marketing. The final chapter provides a general overview of the industry. A brief bibliography is also included.

66. **How to (Try to) Find an Oil Field.** By Doris M. Curtis, et al. Tulsa, Okla.: PennWell Publishing Co., 1981. 94p. illus. $23.95. LC 81-5936. ISBN 0-87814-166-9.

A nontechnical explanation of how geologists explore for oil, this book was written by a committee of geologists from the Houston Geological Society. It includes chapters on the history of oil exploration, exploration tools and methods, geology, leasing, drilling, production, refining, and petroleum supplies. A list of oilfield information centers is provided for persons needing additional general information about the oil industry. An exploration checklist, geologic timetable, and glossary are also included. There are cartoon illustrations.

67. **Introduction to Oil and Gas Technology.** 2nd ed. Edited by Francis A. Giuliano. Boston: IHRDC, 1981. 194p. illus. index. $32.00. LC 81-69432. ISBN 0-934634-48-3.

Readers do not need to be familiar with petroleum terminology or methods to benefit from this book. Written in a basic, easy-to-understand style, *Introduction to Oil and Gas Technology* offers a good overview of petroleum "upstream" functions such as exploration, land work, drilling, production, and transportation. The text is liberally illustrated with diagrams and photographs. Each subject-specific chapter includes a glossary and bibliography. The work is well indexed.

68. **Modern Petroleum: A Basic Primer of the Industry.** 2nd ed. By Bill D. Berger and Kenneth E. Anderson. Tulsa, Okla.: The Petroleum Publishing Co., 1981. 255p. illus. index. $39.95. LC 74-53168. ISBN 0-87814-172-3.

This well-illustrated book does not attempt to discuss the petroleum industry in technical detail. Instead, "it has been designed to give a broad overview of all aspects of the petroleum industry." Its fourteen chapters discuss petroleum history, geology, equipment, drilling, production, storage, transportation, refining, marketing, instrumentation, and petrochemicals. A brief glossary is included. This is a good tool for students, new petroleum company employees, or anyone trying to achieve a basic understanding of petroleum terminology and technology. Also available in a Spanish-language edition, *Petroleo moderno* (ISBN 0-87814-136-7).

69. **Oil and Gas Field Code Master List.** Washington, D.C.: Energy Information Administration; distr., Springfield, Va.: NTIS, 1982- . annual. index. $46.00/yr. (plus $3.00 handling fee). LC 83-644195. ISSN 0738-9809. EIA-0370(84).

Published by the Energy Information Administration (EIA), "this publication provides a standardized code for all identified oil and/or gas fields throughout the United States." Most of the information is compiled from state sources and an effort is made to eliminate duplicate and invalid data from this master list. In the "Field Code Master List," entries are arranged alphabetically by state and then by field name. "Each field name entry contains the field code, field name, geographical information and other related information. The Field Code Index, which follows, is an abbreviated listing sorted by field codes with field names and the state or states in which each field is located." In addition to providing field codes and county codes for recognized oil and gas fields both onshore and offshore, this publication also cross-references unofficial names for officially recognized fields. It is a valuable tool, but requires some practice to gain proficiency in its use.

70. **Oil and Gas Pocket Reference 1984.** 7th ed. Houston, Tex.: National Supply Company, 1984. 58p. free.

This pocket reference is provided free of charge to the public. It compiles facts and statistics on drilling operations, oil and gas exploration, oil and gas reserves, oil and gas production, pipelining, refining, U.S. consumption, imports, and pricing. It also includes a list of common petroleum industry associations, basic energy conversion factors, significant historical dates for the petroleum industry, and a brief glossary. Request copies from National Supply Company, Pocket Reference, 1455 West Loop South, Houston, TX 77027.

71. **Oil & Gas: The Production Story.** By Ron Baker. Austin, Tex.: Petroleum Extension Service of the University of Texas at Austin, 1983. 91p. illus. $8.50. ISBN 0-88698-002-X.

This is another PETEX (Petroleum Extension Service) primer. It is written "for people who have little or no knowledge of what it takes to get oil and gas from their point of origin to the surface where they can be utilized." It serves as an excellent overview and is liberally illustrated with color photographs and drawings. Reference librarians, students, or nonpetroleum professionals will find this book helpful for a basic introduction to well testing and completion, reservoir stimulation, driving mechanisms, wellhead equipment, artificial recovery, storage, and treatment. A glossary of production terms is included at the end of the primer.

72. **Petroleum Engineer's Continuous Tables Fieldbook.** Dallas, Tex.: Energy Publications, 1981. 134p. $14.95.

This handbook contains a selected collection of engineering tables and calculations which have appeared as part of the Continuous Tables series in the journal *Petroleum Engineer International* (ISSN 0164-8322). The data have been grouped into the broad subject categories of drilling engineering, production engineering, reservoir engineering, economics, and general engineering. A table containing metric English unit conversion factors has also been included. Access to specific data is provided through a detailed table of contents. Data are designed for use when a computer or calculator is not handy.

73. **The Petroleum Handbook.** 6th ed. Amsterdam: Elsevier Scientific Publishing Co., 1983. 710p. illus. index. $127.75. LC 83-16409. ISBN 0-444-42118-1.

This handbook provides technical, historical, and economic information as it relates to the petroleum industry. It is directed to people who are new to the industry and need background information on processes and technology. Major areas of discussion include exploration and production, the chemistry of petroleum, refining and marketing, transportation, research, and development. Other topics include natural gas and gas liquids, petrochemicals, unconventional raw materials and synfuels, and environmental considerations. Photographs, charts, and tables illustrate the text. A section on units of measure, a glossary, and a subject index are also provided. Useful to the information specialist, student, and beginning researcher.

74. **The Petroleum Secretary's Handbook.** 2nd ed. By Willene Jackson Lilly. Tulsa, Okla.: PennWell Publishing Co., 1985. 336p. illus. $35.95. LC 84-27384. ISBN 0-87814-278-9.

The text of this handbook is written at a very basic, introductory level. Examples of forms, leases, and contracts are helpful. Each chapter deals with a particular subject specialty for petroleum secretaries such as drilling, production, transportation, and marketing. Chapters include lists of terminology and study guides. It also includes a glossary and a short chapter of "helpful hints."

75. **Practical Petroleum Engineers' Handbook.** 5th ed. By Joseph Zaba and W. T. Doherty. Houston, Tex.: Gulf Publishing Co., 1970. 949p. index. $57.00. LC 58-12306. ISBN 0-87201-744-3.

Although this reference source has not been revised since 1970, it provides valuable data for practicing engineers involved in drilling and production. Some of the subjects covered are general engineering data, steam, tubular goods, drilling, production, and transportation. Several chapters include bibliographies for further reading. A detailed table of contents and index are provided.

76. **Subject Collections: A Guide to Special Book Collections and Subject Emphases as Reported by University, College, Public, and Special Libraries and Museums in the United States and Canada.** 6th ed. By Lee Ash. New York: R. R. Bowker, 1985. 2v. (2196p.). $165.00/set. LC 85-126315. ISBN 0-8352-1967-4 (vol. 1); 0-8352-1968-2 (vol. 2); 0-8352-1917-8 (set).

This guide identifies thousands of special collections in American libraries and museums. Collections may include books, government documents, manuscripts, personal papers, photographs, maps, films, and many other types of materials. Entries are arranged alphabetically by subject and are further arranged by state name. Special collections related to petroleum include petrochemicals, petroleum geology, petroleum engineering, pipelines, industry and trade, history, petroleum law, and legislation.

26 / 5 — HANDBOOKS, MANUALS, AND BASIC TEXTS

Business

77. **Canadian Oil & Gas Handbook.** Edited by Alexandra Worobec. Toronto: Northern Miner Press Ltd., 1981- . annual. $17.00/yr. LC 81-642832. ISSN 0710-622X.

Statistics for Canadian oil and gas production, consumption, wellhead prices, petroleum exports, hydrocarbon reserves, and oil and gas prices can be found in this handbook. The list of Canadian oil and gas companies contains information on assets, debts, operations, production, mining interests, major shareholders, exploration and development, refining, and marketing. Includes maps, table of contents, and glossary, but no index.

78. **Handbook of Energy Technology and Economics.** Edited by Robert A. Meyers. New York: John Wiley & Sons, 1983. 1089p. index. illus. $79.95. LC 82-8477. ISBN 0-471-08209-0.

About one-third of this energy handbook is directly related to the petroleum industry. The most pertinent chapters for petroleum engineers, students, economists, planners, and information specialists are "Petroleum Production," "Petroleum Processing," "Natural Gas," "Oil Shale," and "Oil Sands." Although this book presents many types of data, there is always a strong emphasis on economics. The text is well illustrated and each chapter has an extensive subject bibliography at the end.

79. **Oil Economists' Handbook 1985.** By Gilbert Jenkins. London: Elsevier Applied Science Publishers Ltd.; distr., New York: Elsevier Science Publishing Co., 1985. 378p. $90.00. GB 84-33723. ISBN 0-85334-325-X.

This handbook is meant to be a ready-reference tool — to provide data quickly for oil economists. "There is international coverage of energy resources, energy production, transportation, petroleum refining, petroleum products (including specialised products), storage, energy economics, pricing and energy companies." In addition to the 180 tables of historical statistical data, the handbook includes a glossary of oil industry terminology and a 208-year chronology of events that have had a significant impact on the international energy field.

80. **Oil Property Evaluation.** 2nd ed. By Robert S. Thompson and John D. Wright. Golden, Colo.: Thompson-Wright Associates; distr., Tulsa, Okla.: PennWell Publishing Co., 1985. 1v. (various paging). index. $44.95. LC 84-236256.

This work serves both as a text on basic concepts of oil and gas property evaluation and as a how-to reference source on determining oil and gas prices, costs, working and net reserve interest, taxes, reserve analysis, and forecasting. The authors warn that the sections containing tax information are subject to change. Engineers, accountants, financial analysts, and anyone needing to understand more about how to evaluate oil and gas properties should find this text useful.

81. **Petroleum Measurement Tables.** Philadelphia: American Society for Testing and Materials, 1980. 10v. $270.00/set. LC 80-68070. ISBN 0-89364-021-2 (set).

Originally developed in the late 1940s, these recently revised tables are now used as the international standard for measuring quantities of shipments during custody transfer of bulk petroleum oils and products. This ten-volume set is the result of a cooperative effort by the American Petroleum Institute, the Institute of Petroleum (London) and the American Society for Testing and Materials. "The overall objective of this effort was to meet the worldwide need for a uniform and authoritative publication, based on the most accurate information available. This publication serves as a basis for standardized calculations of measured quantities of petroleum fluids regardless of point of origin, destination, or units of measure used by custom or statute."

Drilling

82. **Air and Gas Drilling Manual.** By William C. Lyons. Houston, Tex.: Gulf Publishing Co., 1984. 183p. index. $35.00. LC 83-12944. ISBN 0-87201-014-7.

This manual, designed for use by engineers in the field, compares air or gas drilling technology with the more popular mud drilling technology. Ten chapters help the engineer to evaluate air/gas drilling operations, perform basic drilling calculations, compare equipment capabilities, and provide solutions to downhole problems. The four appendices include derivations of air and gas volume and pressure requirement equations, calculations of auxiliary examples, conversion tables, and a guide to nomenclature. A detailed subject index is included.

83. **Drilling Data Handbook.** Houston, Tex.: Gulf Publishing Co., 1980. 412p. $55.00. LC 79-56340. ISBN 0-87201-204-2.

This is an English version of the French title *Formulaire du foreur*. It includes primarily graphs and tabular data related to drilling and production work. The information in this handbook is divided into thirteen sections, lettered A-M. Each section begins with a list of the specific tables and information included. Some of the topics covered are physical properties of drill pipe, drill collar, casing, tubing and line pipe; drilling mud; pressure loss; cementing; directional drilling; kick control; fishing; and rotary drilling equipment. Conversion tables are also provided. There is no index.

84. **Drilling Engineering: A Complete Well Planning Approach.** By Neal J. Adams. Tulsa, Okla.: PennWell Publishing Co., 1985. 960p. illus. index. $84.95. LC 84-1110. ISBN 0-87814-265-7.

A theoretical and practical guide to well planning and drilling, written for use by drilling engineers, this work's chapters are arranged by nineteen major topics from well design and planning through drilling and completion. Sample problems and numerous illustrations accompany the text.

85. **Drilling Engineering Handbook.** By Ellis H. Austin. Boston: IHRDC, 1983. 301p. index. $36.00. LC 82-83470. ISBN 0-934634-46-7; 0-934634-54-8(pa.).

This handbook discusses the steps involved in drilling an oilwell. The information is useful to the geologist, drilling engineer, reservoir engineer, and manager. Topics covered include principles of oilwell drilling, drilling fluids, drilling problems, mud logging, drillstem testing, offshore rigs, and drilling in an offshore environment. Three appendices provide sample forms and testing data used in drilling engineering. The text is clearly written and generously illustrated with tables and figures.

86. **Drilling Practices Manual.** 2nd ed. By Preston L. Moore. Tulsa, Okla.: PennWell Publishing Co., 1986. 586p. index. $55.95. LC 85-19086. ISBN 0-87814-292-4.

This textbook covers drilling engineering practices and procedures from making hole to completion. Individual chapters include such topics as rig selection, cost control, hole problems, planning a well, drilling muds, pressure control, directional drilling, and cementing. The text is well illustrated and a detailed index is provided. Each chapter contains a list of references for further reading.

87. **Fundamentals of Drilling.** By John L. Kennedy. Tulsa, Okla.: PennWell Publishing Co., 1982. 216p. index. illus. $43.95. LC 82-13260. ISBN 0-87814-200-2.

This monograph provides basic explanations of drilling techniques, equipment, and associated problems. The overviews are easy to read and are aimed at people not directly involved in the drilling industry. Chapters include information on the history of oil/gas well drilling, rigs, drill bits, drilling fluids, drilling techniques, well control and safety, and well completion. The text is illustrated with photographs, charts, tables, and graphs. It makes good background reading for undergraduates and nontechnical people associated with the drilling industry.

88. **Handbook of Drilling Practices.** By Byron Davenport. Houston, Tex.: Gulf Publishing Co., 1984. 268p. index. illus. $35.00. LC 84-662. ISBN 0-87201-120-8.

"This book is presented not as the last word in drilling, since the technology changes daily, but simply as a guide and reference manual to solve some major problems encountered while drilling oil or gas wells." The focus of the book is on the steps involved in drilling oil and gas wells, from locating the prospect and securing the right to drill to finishing, plugging, or abandoning the well. The text is written in a clear, easy-to-understand format with many illustrations and sample calculations. Both technical and nontechnical people involved in drilling operations may find valuable information in this text. It differs from other handbooks of this kind in that it seeks to bridge the gap between engineers and roughnecks by emphasizing the role each plays in the production process. Oilfield terminology is defined throughout the text and a glossary is also provided for the reader's convenience.

89. **IADC Drilling Manual.** Houston, Tex.: International Association of Drilling Contractors, 1982. 1v. (various paging). illus. $110.00 (IADC members); $165.00 (nonmembers).

Prepared by the Drilling Technology Committee of IADC, this manual includes twenty-one sections discussing various aspects of drilling. Topics include tubular goods, pipe handling equipment, engines, pumps, blow-out preventer equipment, lubrication, mud, hydraulics, and cementing. The text is illustrated with graphs, charts, tables, and figures. Discussions are clear and concise and should be useful to the driller, experienced crewman, or anyone needing explanatory material on drilling operations and equipment.

90. **Mud Equipment Manual.** Houston, Tex.: Gulf Publishing Co., 1985. 1v. (various paging). contact publisher for price. LC 83-161604. ISBN 0-87201-613-7.

Eleven individually authored handbooks have been combined in this one-volume manual. Much of this material was previously presented at International Association of Drilling Contractors meetings between 1975 and 1977. Detailed information is presented on mud systems, degassers, hydrocyclones, mud cleaners and combination separators, centrifuges, agitation and addition, valves, and disposal systems. The manual is well illustrated and is an excellent source for both field and research personnel involved with drilling muds and mud equipment. The index at the end of the first handbook serves as a cumulative index for the entire manual.

5 – HANDBOOKS, MANUALS, AND BASIC TEXTS / 29

91. **Mudfacts Engineering Handbook.** Houston, Tex.: Milchem Incorporated, 1984. 1v. (various paging). contact publisher for price.

Several chapters of this work detail the uses of drilling fluids; specific information on types of fluids and their applications is also included. A large section of engineering data is provided at the end of the handbook. Graphs, tables, and illustrations enhance the text. Available from the Petroleum Extension Service, The University of Texas at Austin, BRC-2, 10100 Burnet Road, Austin, TX 78758.

92. **Pocket Guide for Mud Technology.** Houston, Tex.: IMCO Services, 1981. 188p. contact publisher for price.

Arranged in handbook format, this pocket guide provides information on water-based muds, oil muds, completion fluids, and specific problems related to uses of drilling fluids. The products section contains a chart which compares mud products by trade name. Charts, tables, and graphs are used throughout the handbook to illustrate engineering data. Copies of this publication are available from IMCO, P.O. Box 22605, Houston, TX 77227.

93. **A Primer of Oilwell Drilling.** 4th ed. By Ron Baker. Austin, Tex.: Petroleum Extension Service of the University of Texas at Austin, 1979. 94p. illus. $8.50. LC 81-100379. ISBN 0-88698-080-1.

Anyone looking for an introductory text explaining basic principles in drilling an oil or gas well should examine this primer. Oil and gas terminology is defined within the text as well as in a glossary at the end of the book. Individual sections discuss petroleum personnel, the drill site, rig components, and drilling operations. The text is heavily illustrated. The primer is also available in Arabic-, Chinese-, and Spanish-language editions.

94. **SI Drilling Manual.** Houston, Tex.: Gulf Publishing Co., 1982. 820p. $195.00. LC 82-15466. ISBN 0-87201-211-5.

Each of the twenty-four chapters in this manual deals with a single aspect of the drilling industry. For example, chapter A concerns drilling bits and discusses design, grading, sizes, and handling. The manual was prepared by the Canadian Association of Oilwell Drilling Contractors. Its general format is based on the *IADC Ninth Edition Drilling Manual*. "Although some of the material and basic form remains the same, users of the manual will find much in the way of new data, revised and upgraded specifications and a heavier reliance on the API Specifications." The manual should be used as a guide to locate drilling data, design features, specifications, etc. Includes illustrations, tables, graphs, and photographs.

95. **Volume Requirements for Air & Gas Drilling.** By R. R. Angel. Houston, Tex.: Gulf Publishing Co., c1958, 1980. 94p. $25.00. ISBN 0-87201-890-3.

"This book presents the circulation rates that are required for air and gas drilling. These rates are the minimum necessary to produce velocities in the bottom of the annulus that are equivalent in lifting power to a standard air velocity of 3,000 feet per minute. This standard air velocity is required for best results in drilling dry formations." Data are presented on graphs and entries are arranged in order by size of hole. This handbook is in its third printing.

96. **WACEL Driller's Guide.** Silver Spring, Md.: Washington Area Council for Engineering Laboratories, 1984. 50p. $20.00.

Presented in looseleaf format, this handbook makes it convenient to insert updated ASTM or AASHTO standards as necessary. The guide was prepared to supply basic information to drilling personnel in the Washington, D.C. area, but includes information on drilling and testing procedures, logs, water table conditions, safety, quality control, etc., useful to all drillers. Seven appendices include information on Washington area geology, standard definitions of terms relating to soil and rock, conversion factors, general knowledge related to drilling, an office checklist, and sample boring logs. Drillers, office personnel, and students should consider this guide a good beginning tool but should not rely on it for comprehensive coverage of the drilling field.

Exploration

97. **AGI Data Sheets for Geology in the Field, Laboratory, and Office.** Compiled by R. V. Dietrich, J. T. Dutro, and R. M. Foose. Falls Church, Va.: American Geological Institute, 1982. 150p. $12.95. LC 82-16307. ISBN 0-913312-38-X.

A handy reference tool for anyone involved in the geosciences, this work includes sixty-one data sheets cumulating a variety of reference information. Graphs, charts, tables, and other illustrations are used to summarize the data. A table of contents lists the title of each data sheet. Intended to provide quick reference for field, office, or laboratory personnel.

98. **Coring and Core Analysis Handbook.** By Gene Anderson. Tulsa, Okla.: The Petroleum Publishing Co., 1975. 200p. illus. $39.95. LC 74-33713. ISBN 0-87814-058-1.

This handbook is written in textbook style and provides detailed guidelines for taking a core and analyzing the sample. Although the handbook is ten years old, it is still considered to be a valuable tool for background information and explanations of core procedures and processes. The four appendices provide information on diamond core drilling, core analysis report forms, geologic time table, and conversion factors and useful formulas.

99. **Introduction to Petroleum Geology.** 2nd rev. ed. By G. D. Hobson and E. N. Tiratsoo. Houston, Tex.: Gulf Publishing Co., 1985. 384p. index. illus. $55.00. LC 80-85239. ISBN 0-87201-399-5.

The aim of this text is to familiarize students and nonpetroleum professionals with the basic concepts and theories of petroleum geology. The text is well written and easy to understand. Excellent tables, graphs, figures, and a bibliography enhance the written explanations. The second revised edition has a thirty-two-page supplement summarizing important changes and discoveries that have occurred since the first edition was published in 1981.

100. **Union List of Geologic Field Trip Guidebooks of North America.** 3rd ed. Falls Church, Va.: American Geological Institute, 1978. 253p. index. LC 78-52012. ISBN 0-913312-05-3.

This union list was compiled by the Geoscience Information Society and covers road log, field trip, and geological guidebooks for North America, Bermuda, and the

5 – HANDBOOKS, MANUALS, AND BASIC TEXTS / 31

West Indies. Most guidebooks are arranged by the society or geological survey that sponsored the field trip. One short section on miscellaneous guidebooks is also provided. Each entry lists date of publication, title, and code for library holdings. An alphabetical list of participating libraries, directory of library services, and geographic index are also included. An update to this edition is in preparation.

101. **Well-site Geologist's Handbook.** By Donald McPhater and Brian MacTiernan. Tulsa, Okla.: PennWell Publishing Co., 1983. 77p. illus. $24.95. LC 82-22249. ISBN 0-87814-217-7.

Intended as a working tool for practicing well-site geologists, this source cumulates information on procedures and data gathering techniques used at the well-site. Topics covered include preparation, reporting and notification, supervision and witnessing, and well-site duties. Conversion tables, a list of abbreviations, and sample forms used in reporting are provided in the appendices. It is a very basic handbook, probably most useful to new or inexperienced well-site geologists.

Legal

102. **Arthur Young's Oil and Gas Federal Income Taxation.** 22nd ed. Chicago: Commerce Clearing House, 1984. index. 725p. $30.00. LC 80-82150. ISBN 0-317-19219-1.

This work presents an overview of U.S. tax laws that cover oil and gas operations. Thirty-two chapters provide information on depletion, interests in oil and gas, lease and sale, losses, unitization, partnerships, depreciation, and windfall profits. A table of cases and rulings refers the reader to appropriate chapters and sections in the main body of the text. This book is particularly useful to accountants or tax lawyers dealing with petroleum-related tax matters in the United States. It is also helpful to petroleum information specialists seeking information on a specific oil and gas tax law or act. (This publication continues *Miller's Oil and Gas Federal Income Taxation*.)

103. **The Continental Shelf: Main Acts, Regulations and Guidelines Issued by Norwegian Authorities.** Stavanger, Norway: Norwegian Petroleum Directorate, 1976(?)- . annual. contact publisher for price.

The 1985 edition (ISBN 82-7257-158-7) of this publication provides "a compilation of the main laws, regulations and guidelines" which apply to petroleum-related activities on the Norwegian continental shelf. The text is presented in Norwegian and English. Chapters are arranged by the agency which issues or governs the regulations. Agencies included in *The Continental Shelf* are the Royal Ministry of Social Affairs, the Royal Ministry of Petroleum and Energy, the Royal Ministry of Municipal and Labour Affairs, the Norwegian Petroleum Directorate, the Norwegian Maritime Directorate, the Norwegian Coast Directorate, the Norwegian Telecommunication Administration, the Norwegian Civil Aviation Administration, and the Norwegian State Pollution Control Authority.

32 / 5 – HANDBOOKS, MANUALS, AND BASIC TEXTS

104. **Handbook on Petroleum Land Titles.** By Lewis G. Mosburg, Jr. Oklahoma City, Okla.: The Institute for Energy Development, 1981. index. illus. 353p. $32.00. LC 76-41581. ISBN 0-89419-000-8.

This handbook was written as an introduction "to the general principles which govern the ownership of land in the United States and how this land may be 'leased' for oil and gas purposes." It includes information on ownership and transfer of title to land, ownership and transfer of title to oil and gas, conduct of title examinations, and curing title defects. The appendices are perhaps the most valuable part of the handbook. Many samples of land-related affidavits and forms are included. It is well indexed.

105. **Norwegian Petroleum Guide.** Edited by Helene Borgen Christie. Stavanger, Norway: Stokkand Forlag; distr., Tulsa, Okla.: PennWell Publishing Co., 1984. 259p. $46.95. ISBN 82-90496-04-4.

This guide is a translation of the 1983 *Norwegian Petroleum Guide.* It includes four additional articles not found in the 1983 edition. Each chapter provides a detailed overview of some aspect of the petroleum industry in Norway. Chapters include information on petroleum-related government agencies, legislation, taxation, labor issues, and petroleum technology research.

106. **Petroleum Law Guide 1985.** By W. P. Winston. New York: Elsevier Science Publishing Co., 1985. 146p. index. $37.50. ISBN 0-85334-331-4.

Individuals or companies planning to construct or operate petroleum-related facilities in the United Kingdom may need information about licensing requirements. Although this book was not written by a lawyer, it does contain valuable information related to the necessary authorizations and licenses required for both onshore and offshore petroleum activities. "Part 1 of this book is about law. Law is introduced starting with statute law and leading on to statutory instruments and other forms of law such as codes of practice and licenses." Part 2 explains selected acts of Parliament. Statutory instruments are described in part 3, and part 4 provides information on licensing regulations. This book is not intended for use by students of law, but will be useful to anyone who is trying to locate documents specifying regulations or requirements related to British oil and gas licensing procedures.

107. **Texas Oil and Gas Handbook.** Austin, Tex.: R. W. Byram & Company, 1983. 249p. index. $17.50. LC 81-118493.

This handbook serves as a guide to procedures for filing forms and reports with the Texas Railroad Commission. Operators involved in the production and/or transportation of oil and gas within the state of Texas are required to file certain documents with the state. Copies of the forms and instructions for completing and filing them are included. Additional useful information found in the handbook includes: (1) a listing of required forms for specific situations; (2) a subject index; (3) a ready-reference list showing present form number, previous form number, title of the form, and page reference to the text; (4) Railroad Commission telephone directory; (5) a listing of district offices and their directors; (6) a listing of Texas counties with the appropriate railroad district number; and (7) a county map showing railroad district boundaries.

Offshore

108. **North Sea and North 62 Degrees Atlas: Offshore Oil and Gas Activity and Concession.** Herefordshire, England: Oilfield Publications Ltd., 1984. unpaginated. £25.00.

Locations of oil and gas production for offshore areas of the United Kingdom, Norway, the Netherlands, Denmark, West Germany, and Ireland are shown in this atlas. Detailed maps show field locations by block number and indicate the type of production and locations of pipelines. Oil and gas field names and offshore production installations are listed alphabetically by country in two separate listings. A brief directory of operators and participating companies is also included.

109. **The North Sea Environmental Guide.** Herefordshire, England: Oilfield Publications Ltd., 1984. 68p. maps. index. £35.00.

This unique guide to the environment of the North Sea includes maps on wind distribution, bathymetry, mean sea surface temperature, mean surface air temperature, mean sea level pressure, visibility, cloud cover, precipitation, fog, fifty-year extreme wind speed and wave heights, tidal rise, surface current, and observed wave height. It also contains valuable information on North Sea casualties, installations, field data, and exploration and production companies. A one-page bibliography completes the book. "The degree of variation in North Sea conditions makes project planning and execution very often dependent upon the weather and the physical environment...." This guide should be helpful to anyone involved in planning petroleum operations in the North Sea. Maps are printed on durable, high-quality paper.

110. **The North Sea Platform Guide.** Herefordshire, England: Oilfield Publications Ltd., 1985. 725p. indexes. illus. $280.00.

For the past several years, one of the most difficult tasks for a petroleum information specialist has been to assemble a comprehensive list of offshore structures for any specific geographic area. This guide is the first major effort to provide such a list for the North Sea. Every fixed installation is included as well as "data on those fields for which development plans have reached an advanced stage...." Platforms are listed under the field name and fields are listed alphabetically. The text is liberally illustrated with photographs and artists' conceptions of platforms and related systems. Each platform entry includes information on type, function, location, water depth, contractors, design, fabrication, weight, and additional features. Field entries include basic data on location, license, discovery date, geology, API gravity, reserves, production, and development schemes. This guide is a much needed contribution to the petroleum reference literature.

111. **Offshore Frontiers.** Vol. 5- . Houston, Tex.: Resource Publications, Inc., 1985- . annual. $45.00/yr. LC 84-650518. ISSN 8756-7539.

The 1985 edition of this handbook contains a wealth of information for deepwater offshore activities, divided into four categories—drilling, completions, support, and coatings. The section on drilling contains a listing of wells drilled in water deeper than 2,952 feet, a floating rig survey, and annual wind/wave statistics for selected offshore operating areas. The section on completions contains a completions survey for 1984, a list of abandoned wells, and a list of offshore loading terminals. Remote operating vessels (ROV) are pictured and described in the third section. A listing of ROV manufacturers and contractors is provided. Information on diving support vessels is also included in section 3. The last section, on coatings, contains a survey on coating specifications. The handbook includes a bibliography of the previous year's subsea literature. This work continues *Subsea* (Vols. 2-4).

112. **OREDA Offshore Reliability Data Handbook.** Tulsa, Okla: PennWell Publishing Co., 1984. 386p. $149.95. ISBN 82-515-0087-7.

In this work a joint industry effort was made to compile reliability data "including significant failure modes, failure rate and repair data" for offshore platforms. Both quantitative and qualitative reliability data are presented for safety systems, process systems, electrical systems, utility systems, crane systems, and drilling equipment. An entire chapter is dedicated to a discussion of the handbook's data collection and statistical methods. The "OREDA Handbook should be considered as a first step in a continuing effort to collect data on the reliability of offshore systems."

113. **The Tanker Register.** London: H. Clarkson & Company Ltd., 1960- . annual. £80.00/yr. LC 71-565903. ISSN 0305-179X.

This register provides "a record of all tankers and combined carriers in the world having a deadweight of 10,000 tons and above." It is divided into three main sections. Section 1 lists tankers by name under categories for size and age, giving deadweight for each tanker. Section 2 gives specifics for tankers, arranged alphabetically by name, and includes description, flag, call sign, deadweight, draught, owners or managers, speed, building information, and specifications. The third section is a listing of owners or managers and their tankers. Historical tanker statistics and a brief glossary are also included.

Pipeline

114. **Cost Estimating Manual for Pipelines and Marine Structures.** By John S. Page. Houston, Tex.: Gulf Publishing Co., 1977. 316p. $48.00. LC 76-40868. ISBN 0-87201-157-7.

This manual provides data on "time frames, labor crews and equipment spreads to assist the estimator in capsulizing an estimate for the installation of cross-country pipelines, marshland pipelines, near shore and surf zone pipelines, submerged pipelines, wharfs, jetties, dock facilities, single-point mooring terminals, offshore drilling and production platforms and equipment and appurtenances installed thereon." Data are presented in tabular form. No price estimates are included. This handbook could be useful to pipeline or offshore engineers looking for information on tubular sizes, drilling platform installation times, pipe coating and wrapping materials, etc.

115. **Crude Oil Pipeline Atlas of the United States and Canada.** Tulsa, Okla.: PennWell Publishing Co., 1981. 68p. index. $25.00. LC 82-104642.

This atlas provides detailed maps of oil pipelines in the United States and portions of Canada. Individual pipelines are color-coded and pattern-coded to indicate the operating pipeline company. An alphabetical index to pipeline companies helps relate codes to operators. The atlas also includes a pipeline economic survey and a refining survey. Useful for locating exact routes of pipelines, locations of pipeline pump stations, refinery locations, and pipeline companies.

116. **Handbook of Oceanic Pipeline Computations.** By Alex Marks. Tulsa, Okla.: PennWell Publishing Co., 1980. 524p. index. $83.95. LC 80-20027. ISBN 0-87814-423-7.

5 – HANDBOOKS, MANUALS, AND BASIC TEXTS / 35

This handbook attempts to "consolidate fundamental theoretical oceanic pipeline data for the practicing engineer." It is divided into three main parts plus three appendices. The three main parts cover preliminary pipeline sizing, preliminary corridor selection, and final design. The appendices include offshore petroleum activities and economic factors, flow rate limitations, and conversion factors. The handbook is very helpful to petroleum information specialists as well as practicing engineers, as it includes hydrocarbon fluid characteristics, oceanographic data for selected world oceans, pressure and weight computations, prediction factors for ocean waves, stress analysis, and construction parameters. Several programs for Hewlett Packard HP-67 calculators are also included.

117. **Handbook of Pipeline Engineering Computations.** By Alex Marks. Tulsa, Okla.: The Petroleum Publishing Co., 1979. index. 347p. $83.95. LC 79-127868. ISBN 0-87814-094-8.

This handbook provides calculations useful to the pipeline engineer involved in construction, feasibility studies, and pipeline design. Computations may be done using a Hewlett Packard HP-67 programmable calculator. Step-by-step instructions are provided for performing the computations. References are included for some of the programs. A detailed table of contents and a subject index help the user locate specific applications.

118. **Handbook of Valves, Piping and Pipelines.** By R. H. Waring. Houston, Tex.: Gulf Publishing Co., 1982. 434p. index. illus. $65.00. LC 82-80192. ISBN 0-87201-885-7.

Students, professional engineers, and information specialists will find this handbook an excellent cumulation of data on valves, pipes, and pipelines. The text is generously illustrated with figures and illustrations of specific types of valves and pipes. Many tables, equations, standards, and specifications are included. In addition to descriptions of the actual equipment, services and performance are also discussed. A detailed index is provided.

119. **Pipe Line Rules of Thumb Handbook.** Houston, Tex.: Gulf Publishing Co., 1978. 197p. illus. $23.00. LC 76-52237. ISBN 0-87201-698-6.

This handbook offers quick solutions to everyday pipeline problems. There are six main sections in the handbook: construction, gas, oil and products pipelines, corrosion, economics, and conversion factors. Most data are presented by first posing a question and then working through the answer. Difficult concepts are explained using text and diagrams. The text also uses many graphs and tables to illustrate answers or methods. An excellent office, library, or on-the-job reference book.

120. **Practical Piping Handbook.** By Otto Medel. Tulsa, Okla.: PennWell Publishing Co., 1981. 340p. illus. index. $53.95. LC 81-8553. ISBN 0-87814-169-3.

This handbook presents an overview of practical information on pipes and piping. Many illustrations, charts, tables, and photographs enhance the text. Twenty chapters cover aspects of piping from terminology and standards to field installation and cleaning.

Production

121. **Enhanced Recovery Week EOR Project Sourcebook.** Arlington, Va.: Pasha Publications Inc., 1984. 156p. index. $185.00.

This sourcebook "contains a description of nearly 300 pilot and commercial-scale enhanced oil recovery projects active in the United States as of January 1983." Entries are divided into three major types of enhanced recovery efforts: chemical, miscible, and thermal. These three categories are further subdivided into caustic, microemulsion, polymer, in-situ combustion, and steam injection. Entries in each subdivision are listed alphabetically and include information on field, pay zone, location, operator, contact, project size, number of wells, EOR method, status, and reservoir characteristics. Most entries also include a brief project history and many include cost information. Indexed by oilfield and producing company.

122. **Introduction to Petroleum Production.** By D. R. Skinner. Houston, Tex.: Gulf Publishing Co., 1981. 3v. index. illus. $25.00/vol. 1; $29.00/vol. 2; $25.00/vol. 3. LC 81-6264. ISBN 0-87201-767-2 (vol. 1); 0-87201-768-0 (vol. 2); 0-87201-769-9 (vol. 3).

This set serves as an excellent introductory text and is written for the "semi-technical reader." The first chapter in volume 1 discusses the composition and development of petroleum, while a short section within chapter 3 covers petroleum exploration. The remainder of the volume includes information about the techniques used in the production of petroleum. Illustrations complement the easy-to-understand text. Students, information specialists, and nonengineers will find this text a good starting point in understanding some of the techniques and processes involved in petroleum production. Volume 2 in this three-part series covers production practices, and volume 3 explains facilities, instrumentation, and special topics.

123. **Petroleum Production Handbook.** Edited by Thomas C. Frick. Dallas, Tex.: Society of Petroleum Engineers of AIME, 1962. 2v. index. $34.00/vol. 1; $56.00/vol. 2. LC 60-10601. ISBN 0-89520-206-9 (vol. 1); 0-89520-207-7 (vol. 2).

This 1962 handbook is considered a landmark in the petroleum reference literature. It cumulates basic data for petroleum researchers, field engineers, and others involved in the production of oil and gas. Volume 1 is divided into two sections—mathematics and production equipment. "The Mathematics Section presents the basic tables and calculation procedures required by the person engaged in petroleum production. The Production Equipment Section covers basic types of materials and tools available for use, including their capabilities and proper applications." Volume 2 covers reservoir engineering. "Within the Reservoir Engineering Section are chapters treating formation rocks, fluids and gases, correlation methods, primary and secondary recovery data, and well treating." Many of the individual chapters contain references for further reading. A detailed index provides easy access to the data contained in both volumes. Even though the handbook is more than twenty years old, it is still heavily used for explanations of basic production procedures, equipment, and data. Available from the Society of Petroleum Engineers.

124. **A Primer of Oilwell Service and Workover.** 3rd ed. Austin, Tex.: Petroleum Extension Service of the University of Texas at Austin, 1979. 106p. $6.50. LC 80-154730.

This introductory text provides useful overviews of oilwell service and workover operations. Individual chapters include discussions of well completions, rig equipment, remedial well work, well stimulation, fishing tools and accessories, and economic planning and analysis. The text is generously illustrated and a glossary is provided.

Refining

125. **Organic and Petroleum Chemistry for Nonchemists.** By Louis Schmerling. Tulsa, Okla.: PennWell Publishing Co., 1981. 109p. $24.95. LC 81-10520. ISBN 0-87814-173-1.

Chapters 1-9 of this text contain introductory discussions of organic chemistry and its relationship to the formation of hydrocarbons. "The remainder of this book is devoted to a summary of petroleum chemistry and processes, and definitions (formulas and uses) of the compounds and substances...." An excellent beginning source for people who need a basic understanding of the chemical composition of petroleum and the processes involved in its refining.

126. **Petroleum Processing Handbook.** By William F. Bland and R. L. Davidson. New York: McGraw-Hill Book Co., 1967. 1v. (various paging). index. $88.95. LC 64-66366. ISBN 0-07-005860-1.

Although this handbook has not been updated since 1967, it does provide historical information on petroleum refining processes as well as valuable tables and data. The handbook is divided into fourteen sections and includes a detailed subject index. Diagrams and tables illustrate the text and many sections include references for further reading. Of particular interest are section 3, which discusses different types of processes, section 6, chemicals and catalysts, and section 13, which lists sources of information. It is recommended that this source be used for background information on refining and processing and for a historical perspective on the subject.

127. **Petroleum Refinery Engineering.** 4th ed. By W. L. Nelson. New York: McGraw-Hill Book Co., 1958. 960p. index. $85.00. LC 57-10913. ISBN 0-07-046268-2.

An invaluable text covering petroleum refining principles and processes, this work is useful to chemical engineers, refining personnel, advanced petroleum engineering students, and information specialists. Divided into twenty-four chapters, some of the topics discussed are: the history of refining, composition and physical properties of petroleum, refining equipment, and facility design. Many charts, tables, and graphs illustrate the text. A detailed index is provided.

128. **Petroleum Refining for the Non-technical Person.** By William L. Leffler. Tulsa, Okla.: PennWell Publishing Co., 1985. 190p. index. illus. $39.95. LC 79-87663. ISBN 0-87814-280-0.

This work is intended as an introduction or personal study guide for anyone who has a need to understand some of the basic processes involved in petroleum refining. Some chapters of interest include those on crude oil characteristics, distilling, the chemistry of petroleum, and gasoline blending. There are also individual chapters on processes such as cat cracking, hydrocracking, alklation, isomerization, etc. Illustrations enhance the text and exercises are provided at the end of each chapter for those wishing to check their understanding of the subject.

129. **Technical Data Book—Petroleum Refining.** 4th ed. English ed. Washington, D.C.: American Petroleum Institute, 1985. 3v. $295.00/set.

38 / 5 – HANDBOOKS, MANUALS, AND BASIC TEXTS

Published in looseleaf format, this three-volume manual compiles physical and thermodynamic data and correlations used in petroleum refining processes. Updates are issued periodically. Currently there are fifteen chapters discussing such topics as characterization of hydrocarbons, critical properties, vapor pressure, density, thermal properties, viscosity, diffusivity, and adsorption equilibria. Each chapter contains a bibliography for further reading. This is an important reference tool for researchers and engineers working with hydrocarbons and hydrocarbon mixtures.

Style Manuals

130. **Society of Petroleum Engineers Publications Style Guide.** Richardson, Tex.: Society of Petroleum Engineers of AIME, 1984. 26p. contact publisher for price.

This is a guide for authors who wish to publish materials in any Society of Petroleum Engineers publication. The guide discusses word usage, abbreviations, capitalization, personal titles and names, punctuation, italics, spelling, compounds, numbers, documentation, SI metric conversion factors, standard symbols, and common proofreading notations. This tool is a must for petroleum-related information centers and authors. Available from The Society of Petroleum Engineers of AIME, P.O. Box 833836, Richardson, TX 75083-3836.

131. **Suggestions to Authors of the Reports of the United States Geological Survey.** 6th ed. By Elna E. Bishop and Edwin B. Eckel. Washington, D.C.: Government Printing Office, 1978. 273p. index. contact publisher for price. I19.2:Su 3/5/978.

This manual is intended as a set of guidelines for authors employed by the United States Geological Survey (USGS) in order to insure consistency among their publications. Although some sections pertain solely to USGS documents, others, such as the sections on parts of the report, illustrations, matters of style, and review of English, are of interest to any author publishing geological or petroleum literature.

6 Directories

Directories provide information about people, companies, products, services, training, and associations and are the most common type of reference book available in the petroleum literature. Even though many are published annually, it is not surprising that their information is quickly out-of-date. It is, therefore, wise to supplement statistics and financial data found in directories with other current sources. It is also important to use the most current available edition of any given directory. This chapter focuses on petroleum directories in three broad categories—individuals, companies and other organizations, and products and equipment.

Directories to individuals often include information such as current position, education, age, memberships, and awards. A mailing address and telephone number may also be included. Perhaps the best known directory of petroleum-related people is *Financial Times Who's Who in World Oil and Gas* (entry 136). The biographical entries in this directory include address, current affiliation, and education. Associations' membership directories are also excellent sources of information about people.

If one is trying to locate information on a company or organization, a good place to start is the *U.S.A. Oil Industry Directory* (entry 183) or *The Whole World Oil Directory* (entry 152). These sources list addresses, telephone numbers, telex numbers, and key personnel. Other company directories provide brief company profiles as well. PennWell Publishing Company produces many excellent company-related directories for the petroleum industry. The Midwest Oil Register and Alan Armstrong publishing companies publish series of directories that also list information about petroleum-related companies. By focusing on specific geographic areas, these directories are often able to list smaller companies or division offices of larger companies that are not covered in the PennWell directories.

The third main category of directory, the products and equipment directory, can help locate manufacturers, suppliers, or installers of oilfield equipment. Emphasis here

6 – DIRECTORIES

is on easily identifying sources for selecting and buying equipment, services, or supplies. The *Composite Catalog of Oilfield Equipment & Services* (entry 198) and the *Offshore Contractors & Equipment Directory* (entry 202) are good examples of products and equipment directories.

Annotations in this chapter are arranged alphabetically under the three categories described above. Entries in the category of companies and other organizations are further subdivided into the geographic areas of International, United States, Canada, Europe, Latin America, Middle East, and Southeast Asia.

Individuals

132. **American Association of Petroleum Geologists Annual Report and Membership Directory.** Tulsa, Okla.: American Association of Petroleum Geologists, 1917- . annual. $6.00/yr. (available to members only). ISSN 0149-1423.

Currently published as a special issue of *AAPG Bulletin*, this directory provides a listing of the membership of the American Association of Petroleum Geology. Arranged alphabetically by member's last name, each entry includes name, company affiliation, and address. An indication of AAPG membership category is also included (i.e., active, associate, student, etc.). Following the alphabetical list is a geographical listing. A third section identifies those members who are certified petroleum geologists. Members of the Energy Minerals Division of AAPG are listed in a fourth section. A copy of the Association's annual report is also included.

133. **American Petroleum Institute Directory.** Washington, D.C.: American Petroleum Institute, 1984. 24p. contact publisher for price.

This brief directory lists American Petroleum Institute presidents from 1919 to the present, chairs of the board from 1950 to the present, current officers, members of the Board of Directors, honorary directors, committees of the Board of Directors, API staff, and API office locations. Addresses are provided for all people listed in the directory. Request copies from the American Petroleum Institute, 1220 L St., N.W., Washington, DC 20005.

134. **Directory Interstate Oil Compact Commission and State Oil and Gas Agencies.** By the Interstate Oil Compact Commission. Oklahoma City, Okla.: IOCC, 1984- . irregular. free. LC 84-649711. ISSN 8755-5956.

This book lists the names and addresses of people affiliated with the Interstate Oil Compact Commission (IOCC). The directory is divided into the following listings: governors and their representatives of member states, eleven committees within IOCC and their individual members, and a list of state oil and gas agencies. Although no telephone numbers are provided, this is a good source for determining the names of people to contact concerning topics of interest to the IOCC. Also lists officers and key personnel of the IOCC. Copies are available free from the IOCC, P.O. Box 53127, Oklahoma City, OK 73152, (405) 525-3556.

135. **Directory of North American Geoscientists Engaged in Mathematics, Statistics and Computer Applications.** Lawrence, Kans.: Mathematical Geologists of the United States, 1980(?)- . index. contact publisher for price. LC 80-648093. ISSN 0272-5983.

Mathematical Geologists of the United States is a regional organization affiliated with the International Association of Mathematical Geology. "The directory forms a concise reference to the names, addresses, and professional interests of more than 300 geoscientists." Photographs are included with many of the entries. A specialties index and a geographic index are also provided.

136. *Financial Times* **Who's Who in World Oil and Gas.** London: Longman Group Ltd., 1979- . annual. index. £32.00. LC 83-641802. ISBN 0-582-90313-0 (1982 ed.). ISSN 0141-3236.

This directory provides detailed biographical information about key personnel in the petroleum industry. Worldwide entries are arranged alphabetically by the individual's last name. Personal data, education, current position, and a mailing address are indicated for most entries. An index to organizations serves as a cross-reference by listing the names of individuals alphabetically under their affiliated company or organization name.

137. **Gas Directory and Who's Who.** Kent, England: Benn Business Information Services Ltd., 1975- . annual. £40.00/yr. LC 82-644477. ISBN 0-86382-034-4 (1986 ed.). ISSN 0307-3084.

The 1986 edition of this directory provides thorough coverage of the British gas industry and is divided into the following main sections: "British Gas Corporation"; "Who's Who in the Gas Industry"; "Gas Associations"; "Overseas Contacts"; "Suppliers to the Gas Industry and Who Owns Whom"; "Buyers Guide to Machinery, Equipment, Materials and Services"; and "Approved Domestic Appliances."

Multiple sections provide lists of personnel, addresses, telephone numbers, telex numbers, biographical sketches, product descriptions, and subsidiary information for the British Gas Corporation and other gas-related companies in Great Britain.

138. **The Geological Society of America Membership Directory.** Boulder, Colo.: The Geological Society of America, 1978- . annual. price varies. ISSN 0095-3547.

The membership directory includes information about the society, its members, and its programs. The directory lists name, membership status (fellow, member, or student associate), company, address, telephone number, and year in which the member joined the society. In addition to an alphabetical list of members, there are sections which show membership of the various divisions within the Geological Society of America (GSA), a geographical list of the membership, committees and representatives, and a list of officers of GSA-associated societies. One section lists key personnel at GSA headquarters and their responsibilities. Available to members and to universities or libraries who subscribe to *GSA Bulletin* or *Geology*.

139. **International Association of Drilling Contractors Membership Directory.** Houston, Tex.: International Association of Drilling Contractors, 1985. 189p. index. $40.00 (available to members only).

"The IADC Membership Directory is intended to help you to find the following information about every drilling and well servicing contractor who is a member of the Association: Number and type of rigs operated, areas of operation, addresses, telephone numbers and key personnel. Similar information is also included for producer and associate member firms." In addition, this tool also provides association-related information including publications, officers, committees, and a meetings schedule.

140. **NACE Membership Directory.** Houston, Tex.: National Association of Corrosion Engineers, 1985. 167p. $10.00 (NACE members).

This directory provides an alphabetical list of the membership of the National Association of Corrosion Engineers. Each entry lists member name, address, and company affiliation.

141. **1985 Directory of Certified Petroleum Geologists.** Tulsa, Okla.: American Association of Petroleum Geologists, 1985(?)- . annual. index. $40.00/yr. LC 80-648900. ISSN 0272-1309.

The American Association of Petroleum Geologists certifies those members who have met required standards of education, training, and experience. A description of this certification process and the related requirements is included in this directory. The alphabetical listing of certified petroleum geologists provides name, address, education, and experience for each member. Availability for consulting work is also indicated. A geographic index by country, state, and city is provided.

142. **Society of Petroleum Engineers Annual Technology Review and Membership Directory Issue.** Dallas, Tex.: Society of Petroleum Engineers, 1983- . annual. $25.00/yr. (SPE members); $100.00/yr. (nonmembers). ISSN 0149-2136.

The Society of Petroleum Engineers (SPE) currently publishes the directory of their membership as a special issue of the *Journal of Petroleum Technology* (in May). The directory is an alphabetical list by member name and indicates SPE member grade, year of election to society membership, job title, company affiliation, and address. Other items of interest in the 1985 issue include a review of professional trends, a SPE salary survey, and list of officers for regional sections of SPE. A company index and geographical index are provided as well as a section listing professional services.

Companies and Other Organizations

International

143. **Directory of Oil Well Drilling Contractors.** Tulsa, Okla.: Midwest Oil Register, 1945- . annual. $40.00/yr. LC 58-24722. ISSN 0415-9764.

This worldwide directory provides information on drilling contractors or producers who own rotary or cable tools. Companies are listed alphabetically with listings for Canadian and foreign companies appearing separately after the listing of U.S. companies. Entries include company name, address, telephone number, and key personnel. (Also known as *Midwest Oil Register Oil Well Drilling Contractors*.)

144. **Directory of Pipe Line Companies and Pipe Line Contractors.** Tulsa, Okla.: Midwest Oil Register, 1945- . annual. $25.00/yr. LC 58-35170.

Divided into three main sections, this directory lists pipeline companies operating in the United States, pipeline companies operating outside the United States, and pipeline contractors. Entries in each section are arranged alphabetically by company name and include address, telephone number, description of pipeline activities, and the names of key personnel. The section on U.S. pipeline companies also provides a state geographical index for the companies listed. (Also known as *Midwest Oil Register Directory of Pipe Line Companies and Pipe Line Contractors*.)

145. **Energy Meetings & Trade Shows Worldwide Directory.** Tulsa, Okla.: PennWell Publishing Co., 1984- . irregular. indexes. $45.00. LC 84-640755. ISSN 0742-4337.

Worldwide meetings and seminars of interest to the petroleum, synthetic fuels, and alternate energy industries are listed in this directory. Entries are divided into two sections. Section one includes information on meetings and trade shows for 1984-1991. Section two covers short courses for 1984-1991. All entries are arranged chronologically by meeting date. Information includes conference or course title, location, brief description of content, expected attendance, and information contact. A planning calendar is included. A geographical index and an alphabetical event index are provided. Continues *International Directory of Energy Meetings & Trade Shows.*

146. **Financial Times Oil and Gas International Yearbook.** Essex, England: Longman Group Ltd., 1910- . annual. indexes. £49.00/yr. LC 12-1196. ISBN 0-582-90330-0 (1985 ed.). ISSN 0141-3228.

Even though it's called a yearbook, this publication primarily includes directory information for international oil and gas companies. Companies are divided into two categories: upstream and downstream. Entries are listed in alphabetical order and include business summaries, subsidiary information, property and exploration, development, production, capital, and accounting information as well as address, telephone number, telex number, and list of major company officers. The company listings are followed by a list of brokers and traders and a list of major oil- and gas-related associations. A geographical index and a company index help locate complete entries in the directory section. Other types of summary information included in the yearbook are production, refining and consumption statistics, and a suppliers' directory and buyers' guide. A valuable source for locating annual report information on the world's oil and gas, pipeline, storage, and refining companies.

147. **International Oil Scouts Association Directory.** Austin, Tex.: International Oil Scouts Association, 1985. 152p. $35.00.

This directory includes information on the association, officers, members, code of ethics, and other petroleum-related associations or agencies. Entries for members give an address, telephone number, and corporate affiliation. Entries for other associations or agencies give address, telephone number, and the names of contacts within the organization. Also provides territory maps for North American oil scouts.

148. **Oil Directory of Foreign Companies Outside the U.S.A. and Canada.** Tulsa, Okla.: Midwest Oil Register, 1945- . annual. $20.00/yr. ISSN 0472-7711.

In addition to company name, address, telephone number, and a listing of key personnel, many entries in this directory also include a brief description of company activities and an indication of approximate annual crude and gas production. A separate section lists foreign oil well supply companies. (Also known as *Midwest Oil Register Oil Directory of Foreign Companies Outside the U.S.A. and Canada.*)

149. **Oil-Gas-Marine Directory.** New Orleans, La.: OGM Publishing Co., Inc., 1979- . annual. indexes. $74.50 (1984-1985 ed.). LC 78-645368. ISSN 0162-5675.

This directory is intended to be a comprehensive listing of companies or individuals involved in all phases of the oil, gas, and marine industries. Entries are arranged alphabetically by broad subject headings and include name, address, telephone number, and key personnel. An industry index lists all subject headings used in the directory and provides *see references* for many categories. The company index is a straight alphabetical list of company names with *see references* to advertisements found in the directory. A reference section includes a list of toll-free telephone numbers by subject, a brief glossary, and an index of advertisers. This tool is especially valuable for its subject approach.

150. **Petroleum Engineering Schools Book 1985-86 Academic Year.** (Previous title *Petroleum Engineering and Technology Schools*). Dallas, Tex.: Society of Petroleum Engineers, 1951- . annual. indexes. $17.50/yr.

This reference source provides detailed information on petroleum engineering or petroleum technology curricula at accredited U.S. schools and foreign institutions. It is also helpful for comparing academic programs and identifying potential degree candidates. Entries are divided into two major subdivisions: schools offering petroleum engineering programs and schools offering petroleum technology programs. Information for each entry includes placement director, contact person, accreditation, faculty (names, degrees held, fields of interest, and workload in petroleum engineering), enrollment statistics, program admission requirements, degrees offered and conferred, curriculum description, and the names of student officers and candidates for degrees. A geographic index, an alphabetical index, and a degree program and enrollment index are all provided.

151. **Pipeline Annual Directory and Equipment Guide.** Vol. 25- . Houston, Tex.: Oildom Publishing Co., 1953- . annual. index. $30.00/yr. LC 53-19806.

The 1984-85 issue of this directory includes information on pipeline companies worldwide, pipeline-related engineering services, and pipe coating applicators. Company entries list address, telephone number, corporate ownership, officers, and key personnel. Companies are categorized by major product or service. Two indexes are provided, one arranged by product and the other by company. The directory also includes line pipe tabulations, line pipe mills capabilities, equivalent valve tables, and a pipeline buyers' guide.

152. **The Whole World Oil Directory.** Deerfield, Ill.: Whole World Publishing, Inc., 1979- . annual. 2v. indexes. $103.00/set. LC 81-3468. ISBN 0-938184-11-3 (1985 ed.). ISSN 0148-3609.

The Whole World Oil Directory provides addresses, telephone numbers, and telex numbers for both foreign and domestic companies involved in the oil industry. Entries in the 1985 edition are divided into ten sections: "Oil and Gas Companies – Domestic"; "Oil and Gas Companies – Foreign"; "Drilling Contractors – Domestic"; "Drilling Contractors – Foreign"; "Oil Field Equipment & Pipe and Tubular Products"; "Oil Well Services & Oil Field Specialties"; "Consulting Services"; "Refineries"; "Transmission Pipelines"; and "Marine/Offshore & Transportation." Within each section, entries are alphabetically arranged. Key personnel are often listed. Some entries also list number of employees and give a street address as an alternate for a Post Office box listing. A brief category index precedes each section, with a complete company index and a personnel index following the last section.

153. **Worldwide Petrochemical Directory.** Tulsa, Okla.: PennWell Publishing Co., 1962- . annual. index. $85.00/yr. LC 83-647733. ISSN 0084-2583.

Petrochemical companies throughout the world are represented in this directory. Entries are divided by broad geographic regions: United States, Canada, Latin America, Europe, Africa, the Middle East, and Asia. Each entry includes company name, address, telephone number, telex or cable address, and list of key personnel. Some company entries also include a brief description of services or products. In the 1986 edition, two petrochemical surveys are included at the beginning of the directory. The 1985 Worldwide Petrochemical Survey lists "what feedstocks are processed and what petrochemical products are produced therein." Companies in the survey are listed alphabetically country-by-country. The Worldwide Petrochemical Construction survey identifies "all petrochemical projects under construction on a worldwide basis."

154. **Worldwide Pipeline & Contractors Directory.** Tulsa, Okla.: PennWell Publishing Co., 1976- . annual. index. $50.00/yr. LC 83-648028. ISSN 0146-3349.

This directory provides detailed information for pipeline contractors; U.S. natural gas, crude oil, and products pipeline companies; foreign pipeline companies; and coal slurry pipeline companies. Each entry lists the company name, address, telephone number, telex number, key personnel, and a brief synopsis of company business. Statistics, surveys, and construction reports related to the pipeline industry are also included.

155. **Worldwide Refining and Gas Processing Directory.** Tulsa, Okla.: PennWell Publishing Co., 1942- . annual. index. $85.00/yr. LC 85-10427. ISSN 0277-0962.

Entries in this directory provide company name, address, telephone number, telex number, list of key personnel, and a brief description of the business for worldwide crude oil and natural gas processing companies. Similar information is also given for engineering or construction companies that serve the refining/processing industry. Entries are arranged "alphabetically country-by-country within the large regions of the world." The first 108 pages of the 1986 directory present statistical surveys and construction reports for the refining and gas processing industry. Includes a company name index.

United States

156. **Alaska Petroleum & Industrial Directory.** Anchorage, Alaska: Howell Publishing Co., 1971- . irregular. index. $60.00. LC 72-622422. ISSN 0065-5813.

In the 1985 edition of this directory, fifteen separate sections provide information on more than twenty-five thousand companies operating in Alaska. Entries are not all petroleum-related; there are also sections on fishing, agriculture, mining, culture, travel, and transportation, plus a classified section that subdivides companies into hundreds of specific categories. Each entry includes company name, address, telephone number, a brief statement of services, and the name(s) of key personnel. The directory includes two comprehensive indexes: one for companies and one for personnel. One unique feature of this directory is the "Cardex Professional Directory" of detachable business cards. Approximately 130 Alaskan companies' business cards are included in this section.

157. **Armstrong Oil Directories Eastern United States.** Amarillo, Tex.: Alan Armstrong, 1980- . annual. $50.00/yr.

This Armstrong directory provides an alphabetical listing of companies and individuals involved in the petroleum industry in the eastern United States. Twenty-two states in the northeastern United States are covered by this directory. Information given includes name of company or individual, address, telephone number, and the names of

46 / 6 — DIRECTORIES

key personnel within the organization. This directory is particularly useful in locating small companies and district offices of large oil companies.

158. **Armstrong Oil Directories Gulf Coast.** Amarillo, Tex.: Alan Armstrong, 1980- . annual. $50.00/yr. LC 80-649975. ISSN 0273-4931.

This directory lists alphabetically information on companies and people involved in the petroleum industry in the Gulf Coast area of the United States. This area includes east Texas and the Texas Gulf Coast, Louisiana, Arkansas, Mississippi, Alabama, Georgia, Florida, Tennessee, North Carolina, and South Carolina. Entries give names of individual or company, address, telephone number, and the names of key personnel within the organization. This directory is particularly useful in locating small companies and district offices of large oil companies.

159. **Armstrong Oil Directories Rocky Mountain and Central United States.** Amarillo, Tex.: Alan Armstrong, 1980- . annual. $50.00/yr. ISSN 0273-5229.

This alphabetical listing of oil and gas companies and individuals involved in the petroleum industry covers Kansas, Colorado, Utah, Wyoming, Oklahoma, New Mexico, Montana, North Dakota, South Dakota, Nebraska, Arizona, west Texas, and the Texas Panhandle. Entries include company or individual's name, address, telephone number, and the names of key personnel within the organization. This directory is particularly useful for locating small companies and district offices of large oil companies.

160. **Armstrong Oil Directories Texas and Southeastern New Mexico.** Amarillo, Tex.: Alan Armstrong, 1980- . annual. $50.00/yr.

This volume of the Armstrong series primarily covers Texas and southeastern New Mexico, although some information from other states is also included. Entries list name of company or individual, address, telephone number, and the names of key personnel within the organization. This directory is particularly useful for locating small companies and district offices of large oil companies.

161. **Burmass' Tex-Ok-Kan Oil Directory.** Midland, Tex.: Burmass Publishing Co., 1951- . annual. index. $25.00/yr. LC 84-647096. ISSN 8755-1489.

Production and oilfield service companies in Texas, Oklahoma, and Kansas are listed in this directory. Entries are divided into categories by type of service or activity and list company name, address, and telephone number. Some entries also identify the names of key officers. An alphabetical index and a geographic index are both provided. The listing for oil companies does not appear to be as complete as the listing for the other types of oilfield services.

162. **Directory of Geophysics Education: A Survey of Geophysics in the United States and Canadian Universities and Colleges.** Compiled by William E. Laing. Tulsa, Okla.: Society of Exploration Geophysicists, 1983. 323p. index. $31.25. LC 83-050857. ISBN 0-931830-30-3.

One hundred ninety-two colleges and universities offering geophysics programs are listed in this directory. As the title states, the listings are for colleges and universities in the United States and Canada. Canadian programs are listed first. They appear in alphabetical order by province and then by university or college name. United States listings are arranged alphabetically by state and then by university or college name.

Entries contain the institution address, telephone number, degrees offered, number of students enrolled in the program, a list of faculty members, faculty assistants, course hours required, courses offered, financial aid, and additional comments. Faculty information includes telephone number, degrees held, area of expertise, and related experience. An index by institution is also provided. The foreword to the directory indicates that a second edition will be published in 1986 if warranted.

163. **Directory of Geoscience Departments United States & Canada.** Alexandria, Va.: American Geological Institute, 1952- . annual. indexes. $16.50/yr. LC 83-39001. ISBN 0-913312-73-8 (1984 ed.). ISSN 0364-7811.

Names and addresses of geoscience departments at U.S. and Canadian colleges and universities are listed in this directory. Entries also identify department faculty members, their titles/positions, highest degree held, and field of interest. Schools are listed in two sections—U.S. and Canadian. The U.S. schools are arranged alphabetically by state; Canadian schools are listed alphabetically by the name of the school. A separate index is provided by school name. Other indexes include a specialty index listing faculty members by field of interest and an alphabetical faculty index. One additional section identifies those schools offering field courses or field camps. A useful tool for both students and recruiters.

164. **Directory of Producers and Drilling Contractors: California.** Tulsa, Okla.: Midwest Oil Register, 1945- . annual. $10.00/yr. LC 57-49797.

Identifies oil and gas producers, drilling contractors, and suppliers of rotary tools and cable tools. Entries are arranged alphabetically and include company name, address, telephone number, and key personnel. Many entries also include an indication of approximate annual crude and gas production. A separate section lists consulting engineers and miscellaneous service companies in California. (Also known as *Midwest Oil Register Oil Directory of Producers and Drilling Contractors California*.)

165. **Directory of Producers and Drilling Contractors: Louisiana, Mississippi, Arkansas, Florida, Georgia.** Tulsa, Okla.: Midwest Oil Register, 1945- . annual. $15.00/yr. LC 58-17957.

Most entries in this directory include company name, address, telephone number, listing of key personnel, and type of operation or services provided. Secondary recovery and offshore operations are indicated for many entries. If the listed address is for a branch office, the home office address and telephone number are also provided. A separate listing of consulting engineers and miscellaneous service companies is included at the back of the directory. (Also known as *Midwest Oil Register Directory of Producers and Drilling Contractors Louisiana, Mississippi, Arkansas, Florida, Georgia*.)

166. **Directory of Producers and Drilling Contractors: Oklahoma.** Tulsa, Okla.: Midwest Oil Register, 1945- . annual. $20.00/yr. LC 58-17951.

This directory lists Oklahoma oil and gas producers and drilling contractors. Arrangement is alphabetical by company name. Each entry lists company name, address, and telephone number. Many entries indicate if the company is a producer, a drilling contractor, or a manufacturer of rotary tools and/or cable tools. The approximate annual crude oil and gas production is also listed for many entries along with a description of company operations and key personnel. A separate section includes the names and addresses of consulting engineers in Oklahoma and miscellaneous service companies. (Also known as *Midwest Oil Register Oil Directory of Producers and Drilling Contractors of Oklahoma*.)

167. **Directory of Producers and Drilling Contractors: Rocky Mountain Region, Williston Basin, Four Corners, New Mexico.** Tulsa, Okla.: Midwest Oil Register, 1945- . annual. $15.00/yr. LC 58-17950.

Entries in this Rocky Mountain directory list company name, address, telephone number, a description of company operations, an indication of approximate annual oil and gas production, and the names of key personnel. A separate section lists consulting engineers and miscellaneous service companies in the region. (Also known as *Midwest Oil Register Oil Directory of Producers and Drilling Contractors of Rocky Mountain Region, Williston Basin, Four Corners and New Mexico*.)

168. **Directory of Producers and Drilling Contractors: Texas.** Tulsa, Okla.: Midwest Oil Register, 1945- . annual. $30.00/yr. LC 59-23098.

This directory lists independent producers, drilling contractors, small companies, and consulting engineers operating offices in Texas. Listings for companies which the publisher has been unable to verify have been omitted. Most entries include company name, address, telephone number, key personnel, type of operation, and indications of approximate annual crude production and approximate annual gas production. A separate section lists consulting engineers and miscellaneous service companies operating in Texas. (Also known as *Midwest Oil Register Directory of Producers and Drilling Contractors Texas*.)

169. **Eastern Petroleum Directory.** Trenton, N.J.: Domestic Petroleum Publishers, 1981- . annual. 2v. indexes. $25.00/set.

This two-volume directory lists companies or individuals "involved in the oil and gas industry east of the Mississippi." Volume one is divided into eight industry classifications, including oil and gas companies; individuals; refineries, pipeline, and crude buyers; geophysical contractors; equipment and supplies; services; drilling, workover, and well service contractors; and trade associations. Entries include company name, address, telephone number, and key personnel. Some expanded entries also include information about products, type of service provided, available equipment, etc. Volume two, the geographical directory, contains the same information found in volume one, arranged alphabetically by state, by city, and by company or individual. Volume two includes a cross-reference index to volume one. Both volumes contain company and advertiser indexes.

170. **The Geophysical Directory.** Houston, Tex.: The Geophysical Directory, Inc., 1948- . annual. indexes. $25.00/yr. LC 83-11550.

This directory is useful for identifying companies which provide specific geophysical services and helps locate individual geophysicists worldwide. Entries in the fortieth edition are grouped by type of geophysical service and are arranged alphabetically by company name within each group. Addresses, telephone numbers, telex numbers, and key personnel are given for each entry. Separate indexes for company names, personnel, and advertisers help identify entries within each geophysical service section.

171. **Gulf Coast Oil Directory.** Houston, Tex.: Resource Publications, Inc., 1982- . annual. index. $39.00/yr. LC 83-646787. ISSN 0739-3547.

This directory includes listings of companies and individuals involved in the petroleum industry in Texas, Louisiana, Mississippi, Alabama, and Florida. Entries are divided by type of service and then arranged alphabetically by company name. Each entry includes company name, address, telephone number, telex number, and names of key personnel. An alphabetical index is provided for company and individual names. A products guide identifies the manufacturers and suppliers found in the previous sections. In the 1985 edition, there are twenty-nine companies listed who provide lights for oil derricks.

172. **Houston Oil Directory.** Houston, Tex.: Resource Publications, Inc., 1971- . annual. index. $32.00/yr. LC 83-646318. ISSN 0739-3555.

This directory covers "all phases of the oil industry within a one hundred mile radius of the greater Houston area...." Arranged by type of service, entries include name of company or individual, address, telephone number, and the names and titles of key personnel. An alphabetical index for company and individual names is provided. The section on special services in the 1985 edition includes lists of oil and gas computer services, delivery services, employment agencies, hotels, restaurants, secretarial services, and oil and gas training schools.

173. **Land Drilling & Oilwell Servicing Contractors Directory.** Tulsa, Okla.: PennWell Publishing Co., 1974- . annual. index. $75.00/yr. LC 77-640611. ISSN 0277-0954.

This directory includes directory listings of land drilling and oilwell servicing contractors in the United States and Canada. A few foreign listings are also included. Entries are divided into two listings, U.S. and Canadian, then arranged alphabetically by company name. Foreign contractors are listed in straight alphabetical order by company name. Entries include address, telephone number, and telex number. Many also include a brief company profile and list of key personnel. A geographical cross-reference index allows access to companies by area of operation. An alphabetical company index is also provided.

174. **Midcontinent Petroleum Directory.** Denver, Colo.: Hart Publications, Inc., 1986- . annual. $29.00/yr. ISBN 0-912553-07-3 (1985 ed.).

The Midcontinent region is defined in this directory as Arkansas, eastern Colorado, Kansas, Missouri, Nebraska, Oklahoma, and northern Texas. An effort has been made to list all companies in this region engaged in either the production of oil and gas or services related to the petroleum industry. A section entitled "Government" lists the address and telephone number of each county clerk for the region along with similar directory information for state and federal agencies of interest. Section two contains a selected directory listing of trade associations and professional societies. Company entries include address, telephone number, names of key personnel, and a brief description of business activity. No index is provided for the directory. A single alphabetical company index would make this source easier to use.

175. **The Oil & Gas Directory.** Houston, Tex.: The Oil & Gas Directory, 1970- . annual. indexes. $30.00/yr. ISSN 0471-380X.

This directory provides worldwide coverage of companies and individuals involved in the exploration, drilling, or production of oil and gas. Arranged by broad service categories, entries include the name, address, and telephone number for the company or individual listed. Often the names and positions of key personnel are also indicated. A personnel index and a company index help to locate full entries. The "Company Name Cross Reference" is useful in identifying companies whose names have changed.

176. **Oil Directory of Alaska.** Tulsa, Okla.: Midwest Oil Register, 1945- . annual. $10.00/yr. LC 60-24552. ISSN 0471-3850.

Petroleum-related companies who do business in Alaska are divided into two major sections in this reference work. Section one lists companies engaged in land and lease operations, and section two lists manufacturers and suppliers of oil industry equipment. Each entry includes the company's address, telephone number, and key personnel. (Also known as *Midwest Oil Register Oil Directory of Alaska*.)

177. **Oil Directory of Houston, Texas.** Tulsa, Okla.: Midwest Oil Register, 1945- . annual. $15.00/yr. LC 59-30229. ISSN 0471-3877.

The directory is divided into two listings: a general list of companies in Houston, covering all phases of the oil industry except supply and service, and a listing of manufacturers and suppliers of oil industry equipment. Entries include the company's address, telephone number, key personnel, and a brief description of the business or services. (Also known as *Midwest Oil Register Oil Directory of Houston, Texas*.)

178. **The Oklahoma Petroleum Directory.** Tulsa, Okla.: Oklahoma Petroleum Directory, 1946- . annual. indexes. $18.00/yr. LC 47-34260.

The directory contains an alphabetical listing of companies or people engaged in petroleum exploration or production in Oklahoma. Entries include address, telephone number, and contact person or officers. The 1985 edition has 388 pages of listings. A 288-page classified index, arranged by city, is also provided. "Service and supply companies are listed only in the classified section, unless they are an advertiser."

179. **Pacific Coast Oil Directory.** Brea, Calif.: Petroleum Publishers Inc., 1984- . annual. $35.00/yr.

The 1985 edition of this directory provides information for companies that explore, produce, or refine oil or provide equipment or services to the oil and gas industry on the U.S. Pacific Coast, including Alaska. The directory also includes names, addresses, and telephone numbers for consultants and local associations, agencies, and media.

180. **Permian Basin Oil Directory.** Midland, Tex.: Burmass Publishing Co., 1947- . annual. index. $22.00/yr. LC 55-27973.

Petroleum engineers and people involved in oilfield operations in the Permian Basin will find this directory particularly useful. In addition to company name, address, and phone number, entries often include the names and job titles of key personnel. Companies are grouped by the following main categories: producers; independents and consultants; drilling; well service; special service; specialties; supply companies; pipe, steel, and wire rope; truck and construction; and pipelines, refineries, and plants. This directory offers extensive coverage of petroleum-related services available in the Permian Basin. A substantial portion of the directory is devoted to advertisements.

181. **Rocky Mountain Petroleum Directory.** Denver, Colo.: Hart Publications, Inc., 1955(?)- . annual. $32.00/yr. LC 81-649873. ISBN 0-912553-08-1 (1985 ed.).

This directory is perhaps one of the most complete for the Rocky Mountain region. The 1985 edition is divided into the following sections: companies (exploration, production and land); individuals; pipeliners, refiners, and gas processors; drilling and well servicing; geophysical; equipment and supplies; services; government and associations. The criteria for inclusion are described at the beginning of each section. Entries are listed alphabetically and include name, address, telephone number, type of business operation, and the names and titles of key personnel. The government section includes the address and telephone number for the county clerks in each county in the Rocky Mountain states. Directory information for state and federal agencies is also included in this section. Although no company index is provided for the total directory, it is still a very useful reference tool for individuals doing business in this region.

182. **Southwest Petrodata Report.** Houston, Tex.: Howell Publishing Co., 1985. 4v. (2103p.). indexes. $650.00/set.

This four-volume report includes information on operators, producers, and independents involved in any oil- or gas-related activity in Texas, Louisiana, New Mexico, and Oklahoma. Entries in volumes 1-3 include operator name, address, and telephone number. Entries for some states contain additional information such as state identification numbers, company officers or key personnel, production statistics, and descriptions of available drilling equipment. Data for Texas are the most complete and account for volumes 1 and 2 of the report. Volume 4 contains a company index and a personnel index.

183. **U.S.A. Oil Industry Directory.** Tulsa, Okla.: PennWell Publishing Co., 1962- . annual. indexes. $95.00/yr. LC 72-621148. ISSN 0082-8599.

The 1986 edition of this directory lists company address, telephone number, and telex number for the principle oil companies in the United States. It is divided into several categories: integrated oil companies, independent producers, fund companies, marketing firms, crude oil traders, associations, and government agencies. Entries within each category are arranged alphabetically. Company profiles are included for many entries as well as names and positions of key personnel. Addresses for foreign operations of many U.S.-based companies are also given. A copy of the *Oil & Gas Journal's* "400 Report" is also included. Four indexes are provided: association subjects, government agencies index, a geographic cross-reference index (to companies), and a company index.

Canada

184. **Canadian Oil Industry Directory.** Tulsa, Okla.: PennWell Publishing Co., 1979- . annual. index. $45.00/yr. LC 79-644185. ISBN 0-686-45996-2 (1983 ed.). ISSN 0195-590X.

This directory supplies detailed information about oil companies in Canada. Entries include company name, address, telephone number, telex number, and listing of key personnel. Many entries also include brief company profiles. The directory is divided into ten subject categories. Listings within each category are arranged alphabetically by company name. Subject categories include exploration and production, drilling contractors, oil well servicing contractors, pipeline operators, refining, gas processing, petrochemicals, engineering and construction, associations, and government agencies. A small, but useful, section of surveys and statistics is included at the beginning of the directory.

185. **Directory of Geophysics Education: A Survey of Geophysics in the United States and Canadian Universities and Colleges.** Compiled by William E. Laing. Tulsa, Okla.: Society of Exploration Geophysicists, 1983. 323p. index. $31.25. LC 83-050857. ISBN 0-931830-30-3.

For full annotation see entry 162.

186. **Oil Directory of Canada.** Tulsa, Okla.: Midwest Oil Register, 1900- . annual. $20.00/yr. LC 57-49796. ISSN 0474-0114.

This directory presents an alphabetical listing of oil and gas companies operating in Canada. Entries list company name, address, telephone number, a description of company operations, an indication of approximate oil and gas production, and the names of key personnel. A separate section lists oil well supply companies located in Canada. (Also known as *Midwest Oil Register Oil Directory of Canada*.)

Europe

187. **European Petroleum Directory.** Tulsa, Okla.: PennWell Publishing Co., 1980- . annual. index. $70.00/yr. LC 81-640142. ISSN 0277-0962.

European companies engaged in petroleum-related activities are listed in this directory. A production survey lists names, discovery dates, and production statistics for key European oil and gas fields. Entries are arranged alphabetically, country-by-country, and include company name, address, telephone number, telex number, and cable number. Many entries also include a company profile and the names of key personnel.

188. **European Sources of Scientific and Technical Information.** 5th ed. Detroit, Mich.: Gale Research Co., 1981. 504p. $190.00. LC 80-100666. ISBN 0-582-90108-1.

Information about key European information centers may be obtained from this directory. The first three chapters of the directory list national scientific and technical information centers and patent and standards offices. The remaining twenty-six chapters cover individual subject areas. Arrangement within each chapter is alphabetical by country and then by organization name. "The fullest information given for any organization comprises: title; translation of non-English titles; address, telephone number; telex and cable addresses; affiliation; year of foundation; name of person to whom enquiries should be sent; subject coverage; full information on library facilities and information services; and publications." The title index provides access to the main body by both non-English and English translation of the organization title. Keyword and subject indexes are also provided.

189. **The Geologist's Directory.** 3rd ed. London: The Institution of Geologists, 1980- . biennial. £12.50. ISBN 0-9506906-51 (1985 ed.). ISSN 0260-0463.

This directory provides a convenient guide to geological services, equipment, and information sources in the United Kingdom. Ten main sections include government agencies, education, consultants, special services, buyer's guide, information services, and British companies operating overseas. The final section provides conversion factors, travel information, and other useful data related to geologists in the United Kingdom. Even though the directory is specifically written for members of The Institution of Geologists, geologists worldwide will find the information useful and the format easy to follow.

190. **Guide du pétrole gaz—petrochimie.** Paris: Enercom, 1976(?)- . annual. Fr 510/yr. LC 78-646098. ISSN 0294-1120.

Written entirely in French, this directory provides information on all known oil- and gas-related companies, organizations, and associations active in France. Fifteen sections include information on exploration and production, transportation and storage, industrial installations, refineries, distribution, engineering, suppliers of equipment, education, and key personnel. A review article in the 1985 edition, "Oil in the World," gives a 3-4 page overview of the oil industry in fifteen different countries. Countries are listed alphabetically and entries include a synopsis of exploration, production, refining, consumption, and transportation. Addresses of oil and gas associations and/or operating companies are also listed for each country.

191. **North Sea Oil & Gas Directory.** Surrey, England: Spearhead Publications Ltd., 1972(?)- . annual. indexes. $55.00/yr. LC 83-640042. ISSN 0265-5039.

This directory provides information on companies, organizations, and individuals involved in the exploration and production of oil and gas in the North Sea area. Sections one and four list companies involved in exploration and production and oil industry manufacturers, suppliers, and contractors. Entries include company name, address, telephone number, telex number, a statement of company activity, and a list of key personnel. Section two provides directory information for "government, regulatory and certification authorities, trade associations, learned bodies and research establishments listed alphabetically ..." and divided by country. A brief statement identifies the type of work or service provided by each organization. Both a classified index and a company index are provided in sections 3 and 5. A supplemental update is issued approximately six months after the full directory is published.

192. **Strathclyde Oil Register.** Glasgow, Scotland: Strathclyde Regional Council, 1983- . annual. free. index.

This directory provides addresses, telephone numbers, telex numbers, names of company directors or managers, and a brief statement of products or services for companies that service the oil and gas industry in the Strathclyde Region. A product/service index leads the reader to companies in the directory section. Available free from the Strathclyde Regional Council, 20 India Street, Glasgow G24PF, Scotland.

Latin America

193. **Latin America Petroleum Directory.** Tulsa, Okla.: PennWell Publishing Co., 1971- . annual. index. $45.00/yr. LC 72-613215. ISSN 0193-8738.

Arranged alphabetically, country-by-country, this directory lists petroleum-related companies operating in Latin America. Entries include company name, address, telephone number, telex number, cable number, a brief description of activity, and a listing of key personnel. Additional statistical information is provided in refining, production, petrochemical, and construction surveys. Company and personnel indexes are also provided.

Middle East

194. **Asia-Pacific/Africa-Middle East Petroleum Directory.** Tulsa, Okla.: PennWell Publishing Co., 1984- . annual. index. $70.00/yr. LC 84-6349. ISSN 0748-4089.

Beginning in 1984, PennWell combined two directories, *Asia-Pacific Petroleum Directory* and *Africa-Middle East Petroleum Directory* into this single directory. "The directory is divided into two sections: Asia-Pacific and Africa-Middle East. Directly preceding each section is a map outlining those regions of the world. Immediately following each section map is the appropriate portion of the *Oil & Gas Journal's* Worldwide Production report...." Entries provide addresses, telephone numbers, telex numbers, and a list of key personnel in petroleum-related companies. Some listings also include brief company profiles. Entries are arranged alphabetically by company within each country.

Southeast Asia

195. **China Offshore Oil Directory: A Bilingual Guide to Organizations in the Chinese Offshore Oil Industry.** Hong Kong: Petroleum News Southeast Asia Ltd., 1983. unpaginated. contact publisher for price.

This directory is divided into four main parts. Part 1 lists Chinese petroleum-related companies. Entries are in English and Chinese and include address, telephone number, telex or cable number, name of director, and brief description of services or products. Part 2 lists foreign petroleum-related companies. These companies either maintain offices in Southeast Asia or have interests in the Chinese oil industry. Part 3 is an alphabetical listing, by product name, of products used in the offshore oil industry. Most entries are for companies located in Southeast Asia. Part 4 lists companies providing oil-related services to China. The directory also includes lists of petroleum organizations and a list of personnel.

196. **South East Asia Oil Directory.** Singapore: J. S. Metes & Company (PTE) Ltd., 1977- . annual. $69.00/yr. LC 77-641966. ISSN 0129-4903.

This directory contains one of the most comprehensive listings of information on petroleum-related companies and services in Southeast Asia. The first two sections list directory information for oil companies and supply and service companies. Company name, address, telephone, telex and cable number, key personnel, and a brief description of business activity are provided. The last three sections provide brand name, product, and service indexes to the service and supply companies listed in section 2.

Products and Equipment

197. **Catalogue: British Suppliers to the Oil, Gas, Petrochemical and Process Industries.** London: Energy Industries Council, 1982- . biennial. index. contact publisher for price. LC 82-644950.

This directory is divided into three sections: products and services; names, addresses, and trade names; and product data and advertisements. The first section lists companies under the appropriate product or service offered by that company. The second section lists the companies from section 1 alphabetically and provides the address, telephone number, telex number, and a brief description of business for each entry. In both the products section and the names section, a page number may be listed to the right of an entry. These numbers refer the reader to that company's advertisement in the final section of the directory.

198. **Composite Catalog of Oil Field Equipment & Services.** Houston, Tex.: Gulf Publishing Co., 1929- . biennial. 4v. $1600.00/set.

This four-volume set contains copies of catalogs from companies providing services or equipment to oil- and gas-related industries. Volume 1 contains a classified index to products and services included in all four volumes. Each volume includes a complete company index. Request copies of this catalog from the publisher.

199. **Geoscience Software Directory for the IBM PC & Compatibles.** Boston: IHRDC, 1985. 125p. index. $45.00. ISBN 0-88746-064-X.

This directory contains information on geoscience programs available for use on IBM PCs and IBM-compatible equipment. The first section is alphabetically arranged by the first significant word from the application software title. "Each listing contains known information about the program: program title, price, publisher with address and/or telephone number, brief description, and the hardware requirements." The second section consists of a subject index to section 1. Terms in the subject index are not standardized, however, and this reduces the usefulness of the index.

200. **Hydrocarbon Processing Catalog.** Houston, Tex.: Hydrocarbon Processing Co., 1933(?)- . annual. index. contact publisher for price. LC 65-6066. ISSN 0271-5724.

Copies of catalogs from companies involved in the hydrocarbon processing industry have been collected to form this directory. An alphabetical company index is provided along with a classified index of products and services. Prices, specifications, and descriptions of the products are supplied directly from the companies through their catalogs. Each year's revised catalog replaces the previous issue.

201. **Lloyd's Maritime Directory: The International Shipping & Shipbuilding Directory.** Essex, England: Lloyd's of London Press Ltd., 1982- . annual. index. $80.00/yr. LC 84-642756. ISBN 0-907432-93-X.

Multiple indexes are provided in this directory for vessels, shipowners/managers/agents, shipowners' personnel, and countries. Sections within the directory provide detailed descriptions of individual ships and information about the people and companies who operate them. Further data are provided for towage, salvage and offshore services, shipbuilders and ship repairers, and marine engine builders and repairers. A detailed list of maritime and related organizations is also provided.

202. **Offshore Contractors & Equipment Directory.** Tulsa, Okla.: PennWell Publishing Co., 1969- . annual. index. $85.00/yr. LC 76-640942. ISSN 0475-1310.

This directory changed format with the seventeenth edition (1985). There are no longer two separate volumes for personnel and equipment. Also, no photographs are provided for mobile rigs. The directory still provides information on "contractors and equipment involved in the offshore petroleum industry." The personnel section covers seven types of contractors: offshore drilling and rig owners; workover and well servicing; construction equipment; service, supply, and manufacturing; geophysical companies; diving; and air/marine transportation. Companies are arranged alphabetically within each section. Entries include name, address, telephone number, telex and cable numbers, brief descriptions of organizations, and a list of key personnel.

The equipment section is divided into four sections: mobile drilling rig specifications; tender and fixed platform rig specifications; workover and miscellaneous drilling rig specifications; and construction and pipelay barge specifications. Entries include rig name, owner, varying specifications, and location.

The 1985 edition also includes the "most current version available of *Offshore Magazine*'s Marine Transportation Survey and ROV Operators Survey." A general company index is provided.

203. **The Offshore Drilling Register: A Directory of Mobile Sea-going Rigs.** London: H. Clarkson & Company Ltd., 1975- . annual. £55.00/yr. LC 80-647157. ISBN 0-900291-39-7 (1983 ed.). ISSN 0305-4284.

This register serves as a "record of the larger mobile sea-going rigs in existence, under construction or on order throughout the world. It covers all units capable of operating in at least 50 feet of water and having an overall drilling depth of at least 3,000 feet." It is divided into three main sections. Section 1 lists offshore drilling units by contractor. Contractor address, type of drilling unit, and maximum water depth are included for each entry. Section 2 contains an alphabetical listing of individual drilling units. Extensive information on unit type, contractor, builder, and specifications is given for each entry. The third section gives the same types of information as section 2, for offshore drilling units that are on order or under construction. Entries in this section are grouped by the country where construction is taking place. In addition to these three main sections, the register has a glossary and statistics on mobile offshore drilling rigs worldwide.

204. **The Offshore Service Vessel Register.** London: H. Clarkson & Company Ltd., 1977- . annual. £80.00/yr. ISBN 0-900291-35-4 (1984 ed.). ISSN 0309-040X.

This register is "a record of self-propelled vessels of 100 g.r.t. and over designed, converted or equipped to perform services in connection with the offshore mineral exploration and extraction industry, and of such vessels under construction or on order." Information given for each vessel includes name, type, flag, classification, owner, operator, year built, and specifications. Vessels are listed alphabetically by name. The register also contains a glossary, conversion tables, statistical tables, and an alphabetical list of operators, owners, and their vessels.

205. **Offshore Services + Equipment Directory.** San Diego, Calif.: Greene Dot Inc., 1978(?)- . annual. $35.00/yr.

Twenty-five symbols are used in this publication to represent over four hundred topics related to the oil industry. These symbols and the concepts they represent are defined at the beginning of the directory. Section A, Worldwide Buyer's Guide, provides name, logo, address, telephone and telex numbers, key personnel, and a brief description of companies involved in the offshore industry worldwide. Entries in this section are arranged alphabetically by country and symbols indicate the type of service or equipment category. Section B, "Cross Reference," is arranged by symbol and then by geographic area. Companies are listed alphabetically and listings contain name, address, and telephone number. For more detailed information, the user is referred back to section A. The company information provided in section A is very useful; however the combination of logo and symbols makes the visual appearance of the entries somewhat crowded.

206. **Oil and Oilfield Equipment and Service Companies Worldwide.** Edited by Don Nelson. New York: E. & F. N. Spon, 1985- . irregular. $99.00. LC 84-10580. ISSN 0265-640X.

This first edition of this comprehensive directory presents detailed information on oil-related companies worldwide. The major portion of the directory lists companies in alphabetical order and includes company address, telephone number, telex number, lists of key personnel, and a brief overview of business. Many entries also include ownership and financial data. The directory includes an extensive subject index to equipment, services, and products as well as a personnel, geographic, and company indexes. The format and arrangement of this book make it easy to use as a ready-reference tool.

207. **Petroleum Equipment Directory.** Tulsa, Okla.: Petroleum Equipment Institute, 1955- . annual. index. $15.00/yr.

Directory listings are provided in this work for petroleum equipment manufacturers and distributors. Distributors include companies "which stock, distribute and service equipment in petroleum marketing operations." Distributors are listed alphabetically by state or country. Manufacturers include "companies which manufacture equipment used in petroleum marketing operations" and are listed alphabetically by company name. Each entry includes company name, address, telephone number, year established, key contact person, and a brief description of the company's services or products.

The 1985 edition of the directory includes background information on the Petroleum Equipment Institute (PEI), a list of PEI staff members, a list of PEI affiliates, and product advertisers and member companies indexes.

208. **Petroleum Industry Yellow Pages.** Houston, Tex.: The Whico Atlas Co., 1967- . annual. index. free (contact publisher for copies).

These yellow pages identify products and services used in oil production and transportation worldwide. A total of ten separate publications cover the following regions: "Western U.S. Alaska"; "Canadian"; "Ark-La-Tex, Miss & Ala"; "Mid-Continent & Hugoton"; "East-Midwest"; "Afro-European"; "Gulf Coast"; "Indo-Pacific"; "Permian Basin"; and "American-Latina." Arrangement is by subject with an index provided at the front of each publication. Company name, address, and telephone number are listed. Advertisements are scattered throughout.

209. **Petroleum Software Directory.** Tulsa, Okla.: PennWell Publishing Co., 1984- . annual. indexes. $95.00/yr. LC 85-647725. ISSN 0743-6750.

Arranged alphabetically by software owner and/or distributor, this directory lists petroleum-related computer software. Company address, telephone number, and contact information are provided for each company. Information about software includes program name, applications, program description, language/memory requirements, operating system(s)/special accessories, compatibles, and price information. Two indexes are provided: a company index and a program application index.

210. **Petroleum Training Directory.** Boston: IHRDC, 1984- . annual. $105.00/yr. LC 83-6065. ISSN 0741-8922.

Available on a subscription basis, this publication provides information on training resources for the petroleum industry. The first section of the directory is divided into

58 / 6 — DIRECTORIES

six job categories. Within each job grouping, there are listings of related training courses, audiovisual programs, and selected books. Entries are identified by a unique code. Section 2 lists the training resources from section 1 in three categories: courses, audiovisual, and selected books. Entries include title of the training resource, instructor name(s)/author(s), subject content, and year/date produced or presented. Many of the entries for training courses also provide a location and a price for the course. The third section contains an alphabetical listing of addresses and phone numbers for the companies producing the resources described in section 2. A valuable reference tool for professionals, technicians, and support personnel seeking information about available training for the petroleum industry.

211. **Register of Offshore Units, Submersibles & Diving Systems.** West Sussex, England: Lloyd's Register of Shipping, 1978- . annual. contact publisher for price. ISSN 0141-4143.

This work is divided into the following sections: mobile drilling rigs; submersibles; diving systems; list of support ships, barges, rigs, etc.; work units (ships, barges, and platforms used for offshore work); and owners of offshore equipment. Each section includes information on the name of the offshore unit, the owner, Lloyd's Register classification, and a description of the unit and its operation. The directory section gives address, telephone and telex numbers, and name of units listed in the register. Terms of subscription and prices are available from The General Manager, Lloyd's Register Printing House, Manor Royal, Crawley, West Sussex, RH10 2QN, England.

212. **Undersea Vehicles Directory.** Arlington, Va.: Busby Associates, Inc., 1984- . annual. illus. $60.00/yr.

This directory includes in-depth descriptions of worldwide subsea vehicles, a bibliography of technical papers related to undersea vehicles and underwater nondestructive testing, and directory information for manufacturers or operators of subsea vehicles and instrumentation. Chapters in the 1985 edition are divided into major types of vehicles and vehicles are listed alphabetically within each chapter. A glossary and an index of vehicle specifications are also included.

213. **U.S.A. Oilfield Service, Supply and Manufacturers Directory.** Tulsa, Okla.: PennWell Publishing Co., 1983- . annual. index. $85.00/yr. LC 83-643397. ISSN 0736-038X.

This directory is divided into three categories: oilfield service companies, oilfield supply companies, and oilfield manufacturing companies. Companies are listed alphabetically within each section. Names, addresses, telephone numbers, telex numbers, and cable numbers are provided for each entry. Most entries also include a profile of the company and a listing of key personnel. A company index is included.

7 Statistical Sources

Petroleum statistics appear in a variety of forms and are used primarily for analyses, forecasting, and reporting. It is often difficult to find statistical information in the precise form in which it is needed. Flexibility or interpretation skills may come into play when taking statistics in their existing form and converting them into the necessary format.

The most current statistical data are found in journals or online databases. The reference sources in this chapter either contain actual statistics or guide the reader to other sources of petroleum-related statistics. General reference books that help identify other statistical sources not included in this chapter are Wasserman's *Statistics Sources*[1] and *American Statistics Index*.[2]

214. **Basic Petroleum Data Book: Petroleum Industry Statistics.** Washington, D.C.: American Petroleum Institute, 1981- . 3 issues/yr. $45.00/yr. LC 84-648446. ISSN 0730-5621.

This is an excellent source of current and retrospective worldwide petroleum industry statistics. The data book is divided into fifteen sections: energy, reserves, exploration and drilling, production, financial, prices, demand, refining, imports, exports, offshore, transportation, natural gas, OPEC, and miscellaneous. All tables include the original source of the statistics. Published in January, May, and September. Replaces *Petroleum Data Facts and Figures*.

---

[1]Paul Wasserman, ed., *Statistics Sources* (Detroit, Mich.: Gale Research Co., 1984). LC 84-82356. ISBN 0-8103-0359-0.

[2]*American Statistics Index: A Comprehensive Guide to the Statistical Publications of the U.S. Government* (Bethesda, Md.: Congressional Information Service, 1974- . ISSN 0091-1658).

60 / 7 – STATISTICAL SOURCES

215. **EIA Publications Directory: A User's Guide.** Washington, D.C.: Energy Information Agency; distr., Springfield, Va.: NTIS, 1980- . annual. index. $16.95/yr. DE85009131/XAB. NTIS order no. DOE/EIA-0149(85).

The Energy Information Administration (EIA) of the U.S. Department of Energy has collected and published energy-related statistics since 1977. These statistics include energy production, consumption, prices, resources, and projected supply and demand. The purpose of this directory is to provide access to all EIA publications printed in a given year. The directory includes a detailed subject index, title index, report number index, and complete information on how to order publications. Some of the EIA documents that emphasize petroleum statistics include:

Annual Energy Review

Costs and Indexes for Domestic Oil and Gas Field Equipment and Production Operations

International Energy Annual

Petroleum Marketing Monthly

Petroleum Supply Annual

U.S. Crude Oil, Natural Gas, and Natural Gas Liquids Reserves

Weekly Petroleum Status Report

Students, researchers, and information professionals rely heavily on EIA documents for current and historical statistics presented at a very specific level. This directory and user's guide is available from The National Energy Information Center, EI-20, Energy Information Administration, Forrestal Building, Room IF-048, Washington, DC 20585, (202) 252-8800.

216. **The Energy Decade 1970-1980.** Edited by William L. Liscom. Cambridge, Mass.: Ballinger Publishing Co., 1982. 557p. $125.00. LC 81-22884. ISBN 0-88410-873-2.

This work provides statistical data for major commercial energy sources during the years 1970-1980. Color graphics have been used generously throughout the book. The data have been grouped into eleven chapters in an attempt to illustrate the changing role of energy and its worldwide economic impact. An excellent source for historical statistics on petroleum.

217. **Energy Fact Book.** By E. C. Fox and M. Olszewski. Oak Ridge, Tenn.: Oak Ridge National Lab; distr., Springfield, Va.: NTIS, 1984. 71p. $11.95. DE84016648. NTIS order no. ORNL/PPA-84/3.

This fact book cumulates statistical information on a variety of energy sources. The tables covering petroleum data include information on oil consumption, oil imports, proven reserves, estimated oil production capacity, exploratory oil and natural gas well drilling, and the long-term world oil and gas situation. Petroleum statistics are drawn from the following sources: *Annual Energy Review, Monthly Energy Review, World Energy Industry* (vol. 3, no. 1), *A Desirable Energy Future* (Philadelphia: Franklin Institute Press, 1982), and *The Energy Decade 1970-1980* (entry 216).

218. **Energy Statistics: A Guide to Information Sources.** By Sarojini Balachandran. Detroit, Mich.: Gale Research Co., 1980. 272p. $60.00. LC 80-13338. ISBN 0-8103-1419-3.

This useful reference tool was written by a reference librarian for reference librarians and researchers. Divided into three sections, the guide provides access to specific recurring statistical information found in forty energy journals and serial publications. Section one contains an alphabetical keyword/subject listing. Each entry in this listing identifies the name of the reference source containing the statistic indicated. The sources identified in section 1 are listed alphabetically by title in section 2. Section 3 is arranged by types of energy sources and includes some six hundred additional statistical sources. Most entries in this section, as in section 2, are annotated. A directory of publishers is provided. Author and subject indexes are also included.

219. **Federal Offshore Statistics.** Compiled by Edward P. Essertier. Washington, D.C.: Minerals Management Service, 1984. 123p. OCS Report no. MMS 84-0071.

These statistics provide "a numerical record of what has happened since Congress gave authority to the Secretary of the Interior in 1953 to lease the Federal portion of the Continental Shelf for oil and gas." This publication updates the 1983 edition. "It also extends a statistical series published annually from 1969 until 1981 by the U.S. Geological Survey (USGS) under the title *Outer Continental Shelf Statistics*." The book includes the following types of information for federal offshore land: a five-year leasing schedule, lease sales information since 1954, oil and gas well statistics, pipeline construction data, production and revenue figures, a ranking of offshore production by operator, and reserves data. The report also provides information on oil pollution in the world's oceans, a glossary, and a few conversion factors.

220. **Financial Times Energy World.** London: FT Business Information Ltd., 1985. 144p. $37.00. ISBN 0-903199-88-2.

This publication examines worldwide energy production, consumption, and reserves. Special emphasis is given to the period from 1983 to early 1985. Some of the energy areas covered include oil, natural gas, heavy fuel oil, gasoline, coal, nuclear power, renewable energy, and synfuels. Most chapters highlight a single energy topic and include tables and discussions of activity within specific geographic areas. This is a good source for statistics.

221. **Gas Data Book: Brief Excerpts from Gas Facts.** Arlington, Va.: The American Gas Association, 1977- . annual. 25p. $1.00/yr. ISSN 0433-194X.

This pocket-sized reference provides natural gas statistics extracted from the American Gas Association's *Gas Facts*. See entry 222 for a complete listing of subjects included in both works.

222. **Gas Facts.** Arlington, Va.: The American Gas Association, 1975- . annual. index. $30.00/yr. (nonmembers). LC 72-622849. ISSN 0361-4298.

Gas Facts contains statistical data for the gas utility industry and the gas production industry. Most of the facts found in this handbook are "developed from data contained in the Uniform Statistical Report, a detailed information form prepared annually both by electric and gas utility companies. Information is also derived from reports filed with regulatory commissions, from financial publications, and from statistical reports prepared by government and private agencies." Seventeen chapters provide statistics on reserves, natural gas supply, underground gas storage, distribution and transmission, sales, revenues, customers, prices, consumption, financial issues, construction expenditures, and personnel data. This handbook also includes a glossary of gas industry terminology, conversion tables, a list of gas companies (divided by type of business), and a subject index. An excellent source for retrospective natural gas statistics.

223. **Guide to Petroleum Statistical Information.** New York: American Petroleum Institute, 1983- . annual. index. $50.00/yr. LC 84-641353. ISSN 0742-8464.

Anyone who has searched for petroleum statistical information knows that it appears in many forms (usually not the form one is seeking). This handy guide refers to the original source for over 250 recurring statistical features that are published in the journals covered by the *Petroleum/Energy Business News Index* (entry 22). The 1985 edition of this guide covered thirty journal titles. The guide is divided into three parts. Part A is arranged by journal title and gives an indication of the frequency and type of statistical information covered. Samples of the actual tables described in part A are presented in part B. Part C consists of a subject index. This is an excellent tool for identifying sources for popular petroleum statistics such as crude oil prices, production, pipeline costs, and number of active drilling rigs in the United States.

224. **Joint Association Survey on Drilling Costs.** Washington, D.C.: American Petroleum Institute, 1976- . $15.00/yr. LC 85-11755.

"Contains information on drilling costs derived from a sample of wells, footage, and costs collected from operators of different sizes who drilled wells in various areas of the United States." Costs for oil and gas wells and dry holes in twenty-one states (onshore and offshore) are covered by this survey. Information on those states not individually listed is cumulated in a separate table. The data are presented in tabular form, state-by-state, with an indication of footage cost per depth interval. Summaries of the data and a detailed explanation of how the survey was conducted are included.

225. **Oil and Energy Trends Statistics Review.** Reading, England: Energy Economics Research Ltd., 1970(?)- . annual. £70.00/yr. LC 79-648421.

This review presents statistics and analysis of the oil, gas, and energy industries worldwide. It includes an in-depth calendar of energy-related events for the year. Ten-year spans are given in most of the statistical tables. Tables include reserves, production, geophysical activity, oil and gas exploration/discoveries, active rigs, refining, demand, sales prices, imports, tanker fleet size, and freight rate assessments. A good tool for identifying retrospective statistics.

226. **Oil & Gas Journal Data Book.** Tulsa, Okla.: PennWell Publishing Co., 1985- . annual. index. $24.95/yr. LC 85-394. ISSN 8756-7164.

The *Oil & Gas Journal* is well known for publishing valuable oil and gas statistics and industry surveys. In the second edition of their *Data Book* (1986), selected statistics, reports, indexes, and surveys have been reprinted from the 1985 journal. This yearbook also includes industry indicators such as total U.S. petroleum demand, world crude oil production, international rig counts, etc., from the Oil & Gas Journal Energy Database. The comprehensive index provides access by author, subject, and title, to all articles published in *Oil & Gas Journal* for that year.

227. **The Oil & Gas Producing Industry in Your State.** Washington, D.C.: Independent Petroleum Association of America, 1939- . annual. $5.00/yr. ISSN 0747-2528.

This publication provides petroleum statistical data for thirty-three oil and gas producing states. Individual state sections include a map of the state showing all county/parish divisions with an indication of which counties have oil and gas activity. Some historical data are presented along with an indication of the value of oil and gas, industry statistics, production data, and complete directory information for the state agency or department to contact regarding oil and gas matters. Additional tables compare state statistics on exploration, production, drilling, prices, and employment. Published each September as a special issue of *Petroleum Independent*.

228. **Oil Industry Outlook for the United States.** By Robert J. Beck. Tulsa, Okla.: PennWell Publishing Co., 1983- . annual. illus. $150.00/yr. LC 84-644348. ISSN 0741-3343.

This is a statistics-based outlook on the oil and gas industry, intended to help forecasters and decision-makers understand recent events and trends in the industry. Ten sections cover worldwide supply and demand, U.S. supply and demand, capital expenditures, exploration and production, refining and petrochemicals, transportation, natural gas, and alternate energy sources.

229. **Oil Tanker Databook 1985.** By Michael Champness and Gilbert Jenkins. London: Elsevier Applied Science Publishers Ltd.; distr., New York: Elsevier Science Publishing Co., 1985. 347p. $82.50. ISBN 0-85334-311-X.

This databook presents statistics on the world oil tanker fleet since 1950. Tankers are classified according to size in table 1 and by ownership in table 2. Figures are also included for additions to the tanker fleet, scrapped tankers, combined carriers, LNG and LPG carriers, and freight rates. Other "key" oil statistics such as crude oil prices, world production of crude, total oil consumption, and world refining capacities are given. Conversion factors and a brief glossary are provided. Good source for historical tanker statistics.

230. **OPEC Facts & Figures: A Comparative Statistical Analysis.** Vienna: Organization of the Petroleum Exporting Countries, 1983- . annual. contact publisher for price. LC 83-205363.

These statistics are published in an attempt to educate the user to the role of OPEC and its member countries in worldwide energy economics. "The information has been grouped loosely in four broad sections, beginning with a view of the basic structure of world energy and pattern of its consumption. A short section then follows to highlight OPEC's place in world economic trends. A larger section provides further basic information about the Organization and its Member Countries, with the last section taking a generalized macro view of world energy resources, both hydrocarbon and non-hydrocarbon." Statistics are presented in graphs, charts, maps, and tables. A complete list of statistical sources is also included.

231. **PetroGuide.** London: PetroGuide, 1985. unpaginated. 1985- . quarterly. $550.00/yr.

This statistical handbook presents marketing information on worldwide petroleum production and consumption. Refining capacities are shown for individual countries and oil companies. Data in the producers section show country-by-country the amount of oil being produced and the markets for export. The section on companies is arranged alphabetically by oil company name and shows crude oil supply, product sales data, and refining capacity for each company. *PetroGuide* presents a unique view of market share information within the petroleum industry.

232. **Quarterly Oil and Gas Statistics: Statistiques trimestrielles du pétrole et du gas naturel.** Paris: OECD/IEA, 1984- . quarterly. $107.45/yr. LC 85-648859. ISSN 0378-6536.

The Organization for Economic Co-operation and Development (OECD) and the International Energy Agency (IEA) publish these quarterly oil and gas statistics for the twenty-four member countries of the OECD. This work is intended to "provide rapid, accurate and detailed statistics on oil and gas supply and demand in the OECD area." Data are published approximately three months after they are gathered, making this an up-to-date statistical tool. Three-year historical data are also provided in most cases. Entries are arranged alphabetically by country. Introductions, definitions, and notes are given both in English and in French.

233. **Twentieth Century Petroleum Statistics.** Dallas, Tex.: DeGolyer and MacNaughton, 1945- . annual. illus. index. contact publisher for price. LC 46-25798.

These statistics have been compiled from information published in trade journals, by government agencies, and by the American Petroleum Institute. The bulk of the statistics cover the United States. A few overview statistics such as crude reserves or crude prices are also listed for the world. Each statistical table is accompanied by a color graph or chart which visually represents the tabular data. A comprehensive subject-geographic index is included. An excellent source of very detailed petroleum and energy-related statistics.

234. **Yearbook of World Energy Statistics: Annuaire des statistiques mondiales de l'energie.** New York: Publishing Service United Nations, 1982- . annual. $60.00/yr. LC 85-647150. UN82/17/6, UN.

"A comprehensive collection of international energy statistics prepared by the Statistical Office of the United Nations." Before 1979, this yearbook was called *World Energy Supplies in Selected Years.* Although much of this reference book focuses on general energy data, several in-depth tables present statistics related to hydrocarbon production, refining, trade, and consumption. A good source for historical statistics.

8 Databases

Online database searching is rapidly becoming a preferred method for producing lists of data or bibliographic information on a specific subject, author, or geographic area. Database searching has many advantages over traditional manual methods, including access to large collections of information, the ability to search for combinations of ideas simultaneously, and immediate access to new information as soon as it is available.

This chapter presents descriptions of three types of online, interactive databases that contain information related to the petroleum industry. The first type is the bibliographic database. Bibliographic databases contain references to the published literature in a particular subject area. Textual databases, the second type, usually include abstracts or full text of published materials. The third type of database contains only data or numbers. All three types of databases are accessed via computer and a telecommunications link. Online searching provides the flexibility to combine, revise, or expand search terms while interacting with the database.

The format of this chapter is quite different from that of previous chapters. We have not had the opportunity to assess personally every database included here. Therefore, we have relied on database descriptions supplied by the database producers and vendors. The information is presented in a standardized format to allow for comparison. Each entry lists the following: database name, producer, contact person, brief description of content, format of the information (i.e., bibliographic, data, or textual), time span covered, update frequency, languages, vendor, availability, pricing, downloading restrictions, documentation, and the name of the equivalent printed product.

66 / 8—DATABASES

Some printed works that will help identify additional or future petroleum-related databases are *Directory of Online Databases*,[1] *Computer-Readable Databases: A Directory and Data Sourcebook*,[2] and *DATAPRO Directory of Online Services*.[3]

Bibliographic

235. **Alberta Oil Sands Index (AOSI)**

Producer: Alberta Oil Sands Information Centre, Alberta Research Council
6th Floor, 10010-106 Street
Edmonton, ALB, T5J 3L8, Canada
(403) 427-8382; Telex: 037-2147

Contact Person: Helga Radvanyi-Starr.

Content: The AOSI contains over nine thousand items relating to all aspects of the Alberta oil sands. It provides citations and abstracts after 1970 to the literature on such topics as the region's history, environment, and government; economics and social impact; and geosciences, commercial analysis and properties, separation, upgrading, products, and any related subject.

Format: Bibliographic.

Time Span: From 1900 to the present.

Update Frequency: Biweekly.

Languages: English.

Vendor: Producer.

Availability: Unrestricted.

Price Structure: University of Alberta: $30.00/connect-hour, plus host computer charge; CISTI $45.00/connect-hour plus host computer charge.

Downloading Restrictions: Not supplied.

---

[1]*Directory of Online Databases* (New York: Cuadra/Elsevier, 1979- . quarterly. $95.00/yr. LC 85-648743. ISSN 0193-6840).

[2]Martha E. Williams, ed., *Computer-Readable Databases: A Directory and Data Sourcebook* (Chicago: American Library Association, 1985). 2v. LC 84-18577. ISBN 0-8389-0416-5 (vol. 1); 0-8389-0417-3 (vol. 2); 0-8389-0415-7 (set).

[3]*DATAPRO Directory of Online Services* (Delran, N.J.: DATAPRO Research Corporation, 1985). 1,000p.

Documentation: Alberta Oil Sands Index Online Manual (for University of Alberta's SPIRES system); *Alberta Oil Sands Index-Glossary* (includes a list of subject terms used to index material in the database, a brief explanation of the terms, and some cross-referencing to related terms).

Equivalent Printed Product: Alberta Oil Sands Index.

Information supplied by producer.

236. **APILIT**

   Producer: American Petroleum Institute, Central Abstracting and Indexing Service
   156 William Street
   New York, NY 10038
   (212) 587-9660

   Contact Person: Clara Martinez.

   Content: The APILIT database covers petroleum refining and petroleum products technology, related chemistry and engineering, transportation and storage of petroleum and petroleum products, substitutes for petroleum, related environmental matters, and technology and economics of chemicals used in the oilfield. Sources include journal papers, trade magazine articles, meeting papers, dissertations, letters to the editor, news articles, reprints, technical reports, and translations.

   Format: Bibliographic.

   Time Span: From 1964 to the present.

   Update Frequency: Monthly.

   Languages: English, French, German, Dutch, Japanese, Russian, and others.

   Vendor: System Development Corporation Orbit Search Services
   2500 Colorado Avenue
   Santa Monica, CA 90406
   (213) 453-6194

   Availability: Open only to those users who provide basic support to the API Technical Index. Nonsupporters, however, may access the file for up to three hours per year providing they live in a country where there is a basic supporter.

   Price Structure: Basic supporters of the API Technical Index are charged $60.00/connect-hour, $0.15 per online citation printed, $0.25 per offline citation printed; all others are charged $100.00/connect-hour, $0.25 per online citation printed, $0.30 per offline citation printed.

68 / 8—DATABASES

Downloading Restrictions: Not supplied.

Documentation: Thesaurus available from producer.

Equivalent Printed Product: API Abstracts/Literature.

Information supplied by producer.

237. **APIPAT**

Producer: American Petroleum Institute, Central Abstracting and Indexing Service
156 William Street
New York, NY 10038
(212) 587-9660

Contact Person: Clara Martinez.

Content: The APIPAT database covers patents related to petroleum refining and petroleum products technology, related chemistry and engineering, transportation and storage of petroleum and petroleum products, substitutes for petroleum, related environmental matters, and technology and economics of chemicals used in the oilfield. Patents from Belgium, Canada, France, West Germany, Japan, the Netherlands, South Africa, the United Kingdom, and the United States dating from 1964 are included in the file. Coverage of patents from the U.S.S.R., East Germany, and other countries begins with 1982.

Format: Bibliographic.

Time Span: From 1964 to the present for most countries.

Update Frequency: Monthly.

Languages: English, French, German, Dutch, Japanese, Russian, and others.

Vendor: System Development Corporation Orbit Search Services
2500 Colorado Avenue
Santa Monica, CA 90406
(213) 453-6194

Availability: Open only to those users who provide basic support to the API Technical Index. Nonsupporters, however, may access the file for up to three hours per year providing they live in a country where there is a basic supporter.

Price Structure: Basic supporters of the API Technical Index are charged $60.00/connect-hour, $0.15 per online citation printed, $0.25 per offline citation printed; all others are charged $100.00/connect-hour, $0.25 per online citation printed, $0.30 per offline citation printed.

Downloading Restrictions: Not supplied.

Documentation: Thesaurus available from producer.

Equivalent Printed Product: API Abstracts/Patents.

Information supplied by producer.

238. **ECOMINE**

Producer: Bureau de Recherches Géologiques et Minières, Departement Documentation et Information Géologiques
B.P. 6009
45060 ORLEANS CEDEX, France
(38) 64 34 34
or
Observatoire des Matières Premières
120, rue du Cherche Midi
75006 Paris, France
(1) 556 47 20

Contact Person: Ms. C. Breton, B.R.G.M.

Content: This database contains citations to the world literature on the economics of metallic and nonmetallic minerals, oil and gas, and energy; reserves, new deposits, feasibility, costs, and investments in the mining sector; and production, trade, and legislation.

Format: Bibliographic.

Time Span: From January 1984 to the present. ECOMINE is the continuation of section 221-B, Mineral Economics, included in the databases GEODE (1968-1976) and PASCAL (1977-1983). Both GEODE and PASCAL are available on Telesystemes Questel.

Update Frequency: Monthly.

Languages: French and English are used for descriptors. Abstracts are in French, with 100 percent of the titles given in French and 75 percent given in English.

Vendor: Questel, Inc.
1625 I Street, NW, Suite 719
Washington, DC 20006-3001
(800) 424-9600
or (in Europe)
Telesystemes Questel
83,85 boulevard V. Ouriol
75013 Paris, France
33(1)45 82 64 64

Availability: Request from vendor.

Price Structure: Telesystemes Questel: $65.00/connect-hour, $0.15 per citation printed online, $0.25 per citation printed offline. A 50 percent discount on connect-hours fee is available for subscribers to the weekly press review.

Downloading Restrictions: A special user's contract is available.

Documentation: User's guide.

Equivalent Printed Product: Ecomine, Revue de presse; Ecomine Miniére (weekly press review on mineral economics).

Information supplied by producer.

239. **Energy Bibliography & Index (EBIB)**

Producer: Texas A & M University, Reference Division
College Station, TX 77843-8111
(409) 845-8111

Contact Person: Bill Kinyon.

Content: EBIB indexes nonjournal publications on energy, specifically books, government documents, technical reports, conference proceedings and symposia, maps, pamphlets, and brochures. Major areas of concentration are production, utilization, and conservation of all types of fuel; alternative energy sources; power plants and transmission systems; and economic, political, environmental, and statistical aspects of energy-related issues and activities. Also included are German documents from World War II on synthetic fuel production from coal.

Format: Bibliographic.

Time Span: From 1919 to 1982.

Update Frequency: Approximately biennial.

Languages: English, some German.

Vendor: System Development Corporation Orbit Search Services
2500 Colorado Avenue
Santa Monica, CA 90406
(213) 453-6194

Availability: Unrestricted.

Price Structure: Request from vendor.

Downloading Restrictions: Request from vendor.

8 – DATABASES / 71

Documentation: Request from vendor.

Equivalent Printed Product: Energy Bibliography and Index.

Information supplied by producer.

240. **Energy Data Base (EDB)**

   Producer: Department of Energy, Office of Scientific and Technical
   Information
   P.O. Box 62
   Oak Ridge, TN 37831
   (615) 576-1155

   Contact Person: Dorothy M. Chertok.

   Content: The Energy Data Base (EDB) contains citations, abstracts, and indexes to the world's scientific and technical literature on energy. Its worldwide coverage includes all energy sources (fossil, renewable, and nuclear), conservation, energy policy, advanced energy systems, environmental aspects of energy, and basic scientific studies in the fields of physics, chemistry, biology, and engineering. There are presently 1,400,000 records on EDB.

   Format: Bibliographic.

   Time Span: Although the EDB was started in 1974, it contains material dating back to the 1800s.

   Update Frequency: Semimonthly.

   Languages: English.

   Vendor: Dialog Information Services, Inc.
   3460 Hillview Avenue
   Palo Alto, CA 94304
   (415) 858-2700
      or
   DOE/RECON, Department of Energy, Office of Scientific and
      Technical Information
   P.O. Box 62
   Oak Ridge, TN 37831
   (615) 576-1155
      or
   Mead Data Central
   9333 Springboro Pike
   Dayton, OH 45401
   (513) 865-6800

8 – DATABASES

Availability: Unrestricted. Tapes may be leased from:
National Technical Information Service (NTIS)
U.S. Department of Commerce
Springfield, VA 22161
(703) 487-4807

Searches of EDB are available upon request from:
Western Regional Information Service Center (WRISC)
Lawrence Berkeley Laboratory
Building 50, Room 130
Berkeley, CA 94720
(415) 486-6307

Price Structure: Dialog: $40.00/connect-hour, $0.15 per offline citation printed; Mead, $20.00/connect-hour, $9.00/connect-hour to EDB, $0.02 per line of offline prints; DOE/RECON: $32.00/connect-hour, $0.35 per page of offline prints.

Downloading Restrictions: If the information is going to be used outside the requesting organization (for bibliographies, etc.), EDB should be notified and recognition of the source should be given.

Documentation: The following search aids are available from NTIS: *Energy Data Base: Subject Thesaurus*, November 1984 (DOE/TIC-7000-R6); *Energy Data Base: Subject Thesaurus Permutated Listing*, October 1981 (DOE/TIC-7000-R5(APP)); *Energy Data Base: Subject Categories*, March 1985 (DOE/TIC-4584-R6); *Energy Data Base: Guide to Abstracting and Indexing*, December 1981 (TID-4583-R3); *Energy Information Data Base: Serial Titles*, March 1982 (DOE/TIC-4579-R12); *Energy Information Data Base: Corporate Author Entries*, August 1982 (DOE/TIC-4585-R3); and *Energy Information Data Base: Report Number Codes*, September 1979 (DOE/TIC-85-R13).

Equivalent Printed Product: Energy Abstracts for Policy Analysis, Energy Research Abstracts, Atomindex, Coal Abstracts, and *Biomass Abstracts* as well as the current awareness journals, *Current Energy Patents, Energy and the Environment, Fossil Energy Update, Fusion Energy Update, Solar Energy Update,* and the semimonthly current-awareness bulletin series.

Information supplied by producer.

241. **ENERGYLINE**

Producer: EIC/Intelligence Inc.
48 West 38th St.
New York, NY 10018
(800) 223-6275

Contact Person: Jane Stolarz.

8 – DATABASES / 73

Content: ENERGYLINE includes over seventy-five thousand energy records and provides a primary source for information relating specifically to energy. Data are drawn from many conventional discipline-oriented fields such as chemistry or engineering, but are incorporated into ENERGYLINE only as they relate to energy issues and problems. Coverage includes journals, books, research reports, congressional hearings, conference proceedings, government reports, and statistics. ENERGYLINE provides information on scientific, technical, socio-economic, governmental policy and planning, and current affairs aspects of energy. All documents acquired by EIC for ENERGYLINE are organized into twenty-one categories for quick location according to broad areas of interest.

Format: Bibliographic.

Time Span: From 1971 to the present.

Update Frequency: Ten times/yr.

Languages: English.

Vendor: Dialog Information Services, Inc.
    3460 Hillview Avenue
    Palo Alto, CA 94304
    (415) 858-2700
      or
    ESA-IRS, European Space Agency
    Via Galileo Galilei
    00044 Frascati, Italy
    (06)94011
      or
    System Development Corporation Orbit Search Services
    2500 Colorado Avenue
    Santa Monica, CA 90406
    (213) 820-4111

Availability: Unrestricted; request from vendor.

Price Structure: Dialog & SDC Orbit: $95.00/connect-hour, $0.30 per full record printed offline; ESA-IRS: contact directly for pricing information.

Downloading Restrictions: Not supplied.

Documentation: ENERGYLINE User's Manual.

Equivalent Printed Product: Energy Information Abstracts (since 1976), *Energy Index* (since 1976, 1971-75 corresponds to *Environment Abstracts/Index*).

Information supplied by producer.

242. GEOARCHIVE

Producer: Geosystems
P.O. Box 1024 Westminster
London, SW1P 2JL, England
(01) 222 7305

Contact Person: Valerie Wicks.

Content: GEOARCHIVE covers the entire area of earth science bibliography, placing special emphasis on the economic aspects. Petroleum exploration and production are included in the international coverage of this file. The database contains details of geological maps and in the future will contain more information about these. It will also have the facility to include abstracts and data from the more important papers.

Format: Bibliographic.

Time Span: From 1969 to the present. Coverage for the first five years is incomplete.

Update Frequency: Monthly.

Languages: Publications in any language are covered, but most sources are in English.

Vendor: Dialog Information Services, Inc.
3460 Hillview Avenue
Palo Alto, CA 94304
(415) 858-2700

Availability: Unrestricted.

Price Structure: Per hour connect-time charge; request information from vendor.

Downloading Restrictions: Not supplied.

Documentation: User's guide and thesauri.

Equivalent Printed Product: The following printed publications are all derived from GEOARCHIVE: *Geotitles*, a monthly collection of all the new additions to the database, the *Bibliography of Economic Geology*, which is the bimonthly subset of *Geotitles* having economic implications, and *Geoscience Documentation*, the bimonthly subset containing references to articles of interest to information scientists in the earth sciences.

Information supplied by producer.

243. **GeoRef**

Producer: American Geological Institute, GeoRef Information System
4220 King St.
Alexandria, VA 22302
(800) 336-4764

Contact Person: John Mulvihill.

Content: GeoRef covers worldwide technical literature on geology and geophysics. The database indexes papers from over four thousand serials, and also includes references to books, government documents, maps, American masters' theses, Ph.D. dissertations, conference proceedings, meeting papers and meeting abstracts, field trip guidebooks, and U.S. Geological Survey Open File Reports. About 50 percent of the current citations are from U.S. and Canadian sources, 15 percent from Western Europe, 10 percent from the U.S.S.R., 18 percent from other areas, and 7 percent from international journals.

Format: Bibliographic.

Time Span: 1785 to the present for North American geology; 1967 to the present for geology of other areas.

Update Frequency: Monthly.

Languages: Forty-four languages are indexed.

Vendor: Dialog Information Services, Inc.
3460 Hillview Avenue
Palo Alto, CA 94304
(415) 858-2700
or
System Development Corporation Orbit Search Services
2500 Colorado Avenue
Santa Monica, CA 90406
(213) 453-6194
or
Canada Institute for Scientific & Technical Information
National Research Council
Ottawa, ONT K1A OS2, Canada
(613) 993-1210

Availability: Unrestricted; available from vendor.

Price Structure: $84.00/connect-hour plus $0.30 per online or offline citation on Dialog Information Services, Inc. or SDC Orbit. For-profit organizations spending over $350.00 per calendar year for GeoRef are required by the American Geological Institute to become GeoRef subscribers or to pay a surcharge.

Downloading Restrictions: Unlimited downloading is allowed for purposes of reformatting. Downloading for reuse in private or organizational databases is under review.

Documentation: GeoRef Thesaurus and Guide to Indexing; GeoRef Online Workshop Training Manual; GeoRef Serials List and KWOC Index; User Guide to the Bibliography and Index of Geology; GeoRef Newsletter.

Equivalent Printed Product: Bibliography and Index of Geology.

Information supplied by producer.

244. **Heavy Oil/Enhanced Recovery Index (HERI)**

Producer: Alberta Oil Sands Information Centre, Alberta Research Council
    6th Floor, 10010-106 Street
    Edmonton, ALB T5J 3L8, Canada
    (403) 427-8382; Telex: 037-2147

Contact Person: Helga Radvanyi-Starr.

Content: Oil is found in many forms, from thin, easily flowing liquid to tarry sludge that cannot be pumped out of the ground without special treatment. The articles referenced by this index deal with those sludgy oils and the processes needed to recover them. Anything that deals with the commercial development of heavy oils is included here: economic studies, geosciences, analysis of heavy oils, environmental problems of development, and the methods used to upgrade heavy oils and promote their recovery. Also included in this database are articles on enhanced recovery processes of conventional oil. The citations and abstracts here cover the world's literature since 1900.

Format: Bibliographic.

Time Span: From 1975 to the present; some articles go back to 1900.

Update Frequency: Biweekly.

Languages: English.

Vendor: Producer.

Availability: Unrestricted.

Price Structure: University of Alberta: $30.00/connect-hour plus host computer charge; CISTI $45.00/connect-hour plus host computer charge.

Downloading Restrictions: Not supplied.

Documentation: Heavy Oil/Enhanced Recovery Index Online Manual (for University of Alberta's SPIRES system); *Heavy Oil/Enhanced Recovery Index-Glossary* (includes a list of subject terms used to index material in the database, a brief explanation of the terms, and some cross-referencing to related terms).

Equivalent Printed Product: Heavy Oil/Enhanced Recovery Index.

Information supplied by producer.

245. **OIL**

   Producer: Norwegian Petroleum Directorate (NPD)
   Lagardsveien 80, P.O. Box 600
   N-4001 Stavanger, Norway
   (04) 53 31 60; Telex: 33100 NOPED

   Contact Person: Grete Schanche.

   Content: The OIL database covers Scandinavian petroleum literature presented through patents, journal articles, regulations, government publications, conference papers, books, and newspaper articles. This database carries bibliographic citations to petroleum literature published in Scandinavia. Subjects covered are oil drilling and refining, politics and economics, petroleum exploration and offshore structures, geology and geophysics, pipelining, transportation, storage, and petrochemistry.

   Format: Bibliographic.

   Time Span: From 1974 to the present.

   Update Frequency: Quarterly.

   Languages: Norwegian and English.

   Vendor: Norsk senter for informatikk A/S (NSI)
   Sidsel Aamodt Hoffen, P.O. Box 350 Blindern
   0314 Oslo 3, Norway
   (2) 45 25 08

   Availability: Unrestricted; request password from vendor.

   Price Structure: NOK 480/connect-hour; NOK 300/connect-hour for subscribers to printed equivalent.

   Downloading Restrictions: Not supplied.

   Documentation: Available from vendor.

   Equivalent Printed Product: Request from producer.

   Information supplied by producer.

78 / 8—DATABASES

246. **PASCAL-GEODE**

Producer: Bureau de Recherches Géologiques et Minières,
 Departement Documentation et Information Géologiques
 B.P. 6009
 45060 ORLEANS CEDEX, France
 (38) 64 34 34
 or
 Centre de Documentation Scientifique et Technique, Centre National
 de la Recherche Scientifique
 26, rue Boyer
 75971 PARIS CEDEX 20, France
 (1) 358 35 59

Contact Person: Ms. C. Breton, B.R.G.M.

Content: This database contains citations to the world literature on geology, geophysics, mineral resources, oil and gas exploration, hydrogeology, pollution, marine geology, geochemistry, and other topics related to the earth sciences.

Format: Bibliographic.

Time Span: PASCAL, from 1973 to the present; GEODE, from 1750 to 1976 for French geological literature and from 1968 to 1976 for international literature.

Update Frequency: Monthly.

Languages: French and English descriptors. Abstracts are in French, with 100 percent of the titles given in French and 75 percent in English.

Vendor: ESA/IRS, European Space Agency
 Via Galileo Galilei
 00044 Frascati, Italy
 (06) 94011
 or
 Questel, Inc.
 1625 I Street, NW, Suite 719
 Washington, DC 20006-3001
 (800) 424-9600
 or (in Europe)
 Telesystemes Questel
 83,85 boulevard V. Oriol
 75013 Paris, France
 33(1) 45 82 64 64

Availability: Request from vendor.

Price Structure: ESA/IRS: $47.32/connect-hour, $0.15 per citation printed online, $0.23 per citation printed offline; Telesystemes Questel: (PASCAL)

$65.00/connect-hour, $0.25 per citation printed online, $0.30 per citation printed offline, (GEODE) $65.00/connect-hour, $0.15 per citation printed online, $0.25 per citation printed offline.

Downloading Restrictions: A special user contract is available.

Documentation: Lexique Français-anglais des sciences de la Terre; user's guide (in French).

Equivalent Printed Product: PASCAL-THEMA Sciences de la Terre, Section T022; *PASCAL-FOLIO*, Sections F40-F47.

Information supplied by producer.

247. **P/E News**

   Producer: American Petroleum Institute, Central Abstracting and Indexing Service
   156 William Street
   New York, NY 10038
   (212) 587-9660

   Contact Person: Clara Martinez.

   Content: The P/E News database covers political, social, and economic news related to the energy industries. News items are taken from articles, contract award announcements, book reviews, errata, interviews, letters to the editor, news articles, obituaries, columns, statistics, photo captions, and editorials.

   Format: Bibliographic.

   Time Span: From 1975 to the present.

   Update Frequency: Weekly.

   Languages: English and French.

   Vendor: Dialog Information Services, Inc.
   3460 Hillview Avenue
   Palo Alto, CA 94304
   (415) 858-2700
      or
   System Development Corporation Orbit Search Services
   2500 Colorado Avenue
   Santa Monica, CA 90406
   (213) 453-6194

   Availability: Unrestricted.

80 / 8 – DATABASES

Price Structure: Dialog: $96.00/connect-hour, $0.25 per full record printed online, $0.35 per full record printed offline; SDC Orbit: $95.00/connect-hour, $0.25 per online citation printed, $0.30 per offline citation printed. Basic supporters of the API Technical Index are charged $0.15 per online citation printed and $0.20 per offline citation printed.

Downloading Restrictions: Not supplied.

Documentation: An index guide and keyword list is available from the producer.

Equivalent Printed Product: None.

Information supplied by producer.

248. **TULSA**

Producer: Petroleum Abstracts, The University of Tulsa,
        Information Services Division
        600 S. College
        Tulsa, OK 74104
        (918) 592-6000, ext. 2296; Telex: 49-7543 INFOSVC TU TUL

Contact Person: Janis Mattinson.

Content: The TULSA database contains bibliographic references to scientific articles, patents, books, and government documents on the following subjects as they apply to the exploration and production of oil and gas: geology, geochemistry, geophysics, drilling, well logging, well completion and servicing, reservoir studies, recovery methods, transportation and storage, corrosion, ecology and pollution, alternate fuels and energy, and supplementary technology.

Format: Bibliographic.

Time Span: From 1965 to the present.

Update Frequency: Weekly.

Languages: Primarily English; also French, German, Russian, Chinese, Spanish, Portuguese, Japanese, Norwegian, Polish, Hungarian, Dutch, Czechoslovakian, and Serbo-Croatian (with English titles and abstracts).

Vendor: System Development Corporation Orbit Search Services
        2500 Colorado Avenue
        Santa Monica, CA 90406
        (213) 453-6194

Availability: Subscription; available from producer.

Price Structure: Rates vary according to type of license—subscription rates based on total company assets relating to the petroleum industry.

Downloading Restrictions: Arrange with producer.

Documentation: Exploration and Production Thesaurus; Geographic Thesaurus; Alphabetic Subject Index; Description Frequency List; KWOC (Keyword Out of Context) List.

Equivalent Printed Product: Petroleum Abstracts.

Information supplied by producer.

Data/Textual

249. **AAPG/CDS (American Association of Petroleum Geologists/Committee on Statistics of Drilling) Exploratory Well File**

    Producer: University of Oklahoma, Information Systems Programs,
    Energy Resources Institute
    P.O. Box 3030
    Norman, OK 73070
    (405) 325-1600

    Contact Person: Charlotte Knowles, consultation coordinator.

    Content: Basic information pertaining to wells drilled in the United States is reported by means of an individual well ticket initiated by respondents to the American Petroleum Institute (API), where data are collected and maintained. The University of Oklahoma is the repository for the data. Exploratory well classifications include new field wildcat, new pool wildcat, deeper pool test, shallower pool test, and outpost (extension) test. The types of information in a record are the API well number, completion date, depths, well classifications, number of completions, operator identification, Section-Township-Range (where available), AAPG geological province, deepest formation, code, name, pay code and pay name, field name, and remarks. A particularly useful label in the Exploratory Well File contains information relating to the estimated ultimate yield of successful exploratory wells. These data are represented on each well record by an alphabetic code which corresponds to one of several categories of expected yield. The estimated yields for these categories are expressed in millions of barrels of oil and billions of cubic feet of gas. The database is indexed by state, basin code, regulatory district, initial and final Lahee class, year reported, and well classification.

    Format: Data/textual.

    Time Span: From 1954 to 1983.

82 / 8 — DATABASES

Update Frequency: Annual.

Languages: English.

Vendor: Producer.

Availability: Restricted. This offline database can be leased for installation on a company's in-house computer. Printouts of magnetic tapes containing partial or summary data are available from the University of Oklahoma.

Price Structure: Not supplied.

Downloading Restrictions: Not supplied.

Documentation: Not supplied.

Equivalent Printed Product: Not supplied.

Information supplied by producer.

250. **API Master Well File**

Producer: University of Oklahoma, Information Systems Programs,
    Energy Resources Institute
    P.O. Box 3030
    Norman, OK 73070
    (405) 325-1600

Contact Person: Charlotte Knowles, consultation coordinator.

Content: Basic information pertaining to wells drilled in the United States is reported by means of an individual well ticket initiated by respondents to the American Petroleum Institute (API), where data are collected and maintained. The University of Oklahoma is the repository for the data. The API Master Well File includes information on exploratory wells, development wells, stratigraphic tests, service wells, oil wells drilled deeper (exploratory), and oil wells drilled deeper (development). Contained in a record are the API well number, completion date, depth, well classification, number of completions, operator identification, Section-Township-Range (where available), American Association of Petroleum Geologists geological province, state, and county. The database is indexed by state, basin code, regulatory district, year reported, and well classification.

Format: Data/textual.

Time Span: From 1966 to 1983.

Update Frequency: Annual.

Languages: English.

Vendor: Producer.

Availability: Restricted. This offline database can be leased for installation on a company's in-house computer. Printouts of magnetic tapes containing partial or summary data are available from the University of Oklahoma.

Price Structure: Not supplied.

Downloading Restrictions: Not supplied.

Documentation: Not supplied.

Equivalent Printed Product: Not supplied.

Information supplied by producer.

251. **ARGREP (Petroleum Argus Daily Market Report)**

Producer: Petroleum Argus, Star House
104-108 Grafton Rd.
London NW5 1BD, England
(01) 485-8792; Telex: 21277 ARGUS G

Contact Person: Not supplied.

Content: Daily reports on spot prices for crude oil and petroleum products in leading European centers are included in this database. Trends and quoted prices for the northwest European and west Mediterranean markets cover gasoil, motor gasoline, naptha, jet kerosene, and high- and low-sulphur heavy fuel oils. The database provides bid-ask ranges of both F.O.B. (free on board) and C.I.F. (cost, insurance, and freight) prices. It also includes price, volume, and variety of crude oils traded in the western European market.

Format: Data/textual.

Time Span: The most recent 50 reports are available.

Update Frequency: Daily.

Languages: English.

Vendor: I. P. Sharp Associates
2 First Canadian Place, Suite 1900
Toronto, ONT M5X 1E3, Canada
(416) 364-5361, (800) 387-1588; Telex: 0622259

Availability: Subscription; available from producer.

84 / 8 — DATABASES

Price Structure: Not supplied.

Downloading Restrictions: Not supplied.

Documentation: I. P. Sharp Energy Reference Guide.

Equivalent Printed Product: Petroleum Argus.

Information supplied by vendor.

252. **CMAI (Chemical Market Associates Petrochemical Market Reports)**

Producer: Chemical Market Associates, Inc.
11757 Katy Freeway, Suite 750
Houston, TX 77079
(713) 531-4660; Telex: 792318

Contact Person: Not supplied.

Content: CMAI includes three analytical petrochemical reports that are issued monthly. Each focuses upon a critical dimension of the petrochemical marketplace and provides textual commentary on current developments and future prospects as well as numeric forecasts of emerging trends. Reports vary in length, averaging five to seven pages of text plus two to four pages of tables. The C4 Market Report concentrates primarily on butadiene, butylenes, and their derivatives, while the Monomers Market Report analyzes developments for ethylene, propylene, styrene, vinyl chloride, and their derivatives. The Fibers Intermediate Market Report focuses on polyester, nylon, and acrylic fiber intermediates.

Format: Data/textual.

Time Span: The two latest issues of each report are available.

Update Frequency: Semimonthly or monthly.

Languages: English.

Vendor: I. P. Sharp Associates
2 First Canadian Place, Suite 1900
Toronto, ONT M5X 1E3, Canada
(416) 364-5361, (800) 387-1588; Telex: 0622259

Availability: Subscription; available from producer.

Price Structure: Not supplied.

Downloading Restrictions: Not supplied.

Documentation: I. P. Sharp Energy Reference Guide.

Equivalent Printed Product: CMAI newsletters.

Information supplied by vendor.

253. **DeWitt (DeWitt Petrochemical Newsletters)**

    Producer: DeWitt and Company
    3650 Dresser Tower
    601 Jefferson St.
    Houston, TX 77002
    (713) 652-0576; Telex: 762854 HOU

    Contact Person: Not supplied.

    Content: This database contains weekly analytical reports, price quotations, and forecasts for a broad range of petrochemical products. Sixteen distinct newsletters and forecasts are included in the coverage of the database.

    Format: Data/textual.

    Time Span: Up to 3 latest issues of each newsletter and one issue of each forecast are available.

    Update Frequency: Newsletters, weekly; forecasts, monthly.

    Languages: English.

    Vendor: I. P. Sharp Associates
    2 First Canadian Place, Suite 1900
    Toronto, ONT M5X 1E3, Canada
    (416) 364-5361, (800) 387-1588; Telex: 0622259

    Availability: Subscription; available from producer.

    Price Structure: Not supplied.

    Downloading Restrictions: Not supplied.

    Documentation: I. P. Sharp Reference Guide.

    Equivalent Printed Product: DeWitt newsletters.

    Information supplied by vendor.

254. **Drilling Information Services (DIS)**

    Producer: Adams and Rountree Technology, Inc.
    4801 Woodway, Suite 260 East
    Houston, TX 77056
    (713) 961-1918

Contact Person: Betty Quay.

Content: The Drilling Information Services database contains historical drilling information. DIS includes information from bit records, mud records, log headers, and log resistivities (where appropriate). Primary coverage is for Louisiana, Texas, Oklahoma, and offshore Gulf of Mexico.

Format: Data/textual. Text information is kept in the form of well name, lease name, operator, etc., and in the form of daily comments concerning drilling progress.

Time Span: Approximately 1965 to the present.

Update Frequency: Daily, per request from users for information in a particular area.

Languages: English.

Vendor: Producer.

Availability: Subscription. Access to the database is provided through General Electric Information Services Co. (GEISCO).

Price Structure: A one-time, corporate-wide, subscription fee of $1500.00 must be paid to Adams and Rountree Technology, or 1500 records can be contributed to the database. (A&RT assumes responsibility for copying and verifying the records.) A detailed description of charges assessed during searching of the database is available from the producer. "Unlimited usage rates" are available from A&RT (based on a one-year commitment). In addition to these charges, GEISCO requires a minimum monthly charge for use of their services.

Downloading Restrictions: None.

Documentation: User's guide.

Equivalent Printed Product: None.

Information supplied by producer.

255. **Dwight's Petroleum Data System (DPDS)**

Producer: Dwight's Energydata, Inc., Dwight's PDS Services (DPDS)
    4350 Will Rogers Parkway, Suite 101
    Oklahoma City, OK 73108
    (405) 942-7713

Contact Person: Customer support.

Content: There are seven databases in the Dwight's Petroleum Data System. The primary database, TOTL, contains data on oil/gas fields and/or reservoirs of the United States and Alberta, Canada. Other databases in the system contain information on federal offshore bidding activity, lease and ownership status, enhanced recovery projects, brine, crude oil and gas samples, and Canadian historical production and field/reservoir data.

Format: Data/textual.

Time Span: TOTL contains data for fields and reservoirs with discovery years dating from the late 1800s. Annual production data dates from 1968 to the current year; cumulative production dates from the year of discovery.

Update Frequency: Up to three updates per year for TOTL; updates to other files as information becomes available.

Languages: English.

Vendor: Producer.

Availability: Available for in-house installations through a commercial time-sharing service, and by direct consultation service.

Price Structure: Request from producer.

Downloading Restrictions: Certain copyright, use, and warranty restrictions apply; request from producer.

Documentation: User's guide, database inventories, and pocket references.

Equivalent Printed Product: None.

Information supplied by producer.

256. **Electronic Rig Stats (ERS)**

Producer: PennWell Publishing Co.
P.O. Box 1260
Tulsa, OK 74101
(918) 832-9245

Contact Person: Larry Lofton.

Content: Electronic Rigs Stats includes data on all mobile rigs worldwide, with thirty-five pieces of information about each rig. The ERS database is available twenty-four hours a day. Types of data available include: name, owner, former name, former owner, rig management, type, design, rated water depth, rated drilling depth, dynamic positioning, operator, location, concession and well designation, spud date, water depth at wellsite, target depth of well, next location, status, contract information, date rig available, derrick, drawworks, mudpumps, power source, blowout preventors, cranes, cementing unit, logging services, quarters, year rig built, fabricator, delivery variable, deck load, deadweight tonnage, dimensions of hull, registry, and mishaps.

Format: Data/textual.

Time Span: Current data.

Update Frequency: Three times/wk.

Languages: English.

Vendor: Producer.

Availability: Subscription.

Price Structure: Annual subscription fee $1,575.00; connect-time charge $90.00/hour.

Downloading Restrictions: None.

Documentation: Training guide, user's manual.

Equivalent Printed Product: Worldwide Offshore Rigfinder (partial information only).

Information supplied by producer.

257. **ICIS (Independent Chemical Information Services)**

Producer: Independent Chemical Information Services, La Tour Grand House
28 avenue Alfred Belmontet
92210 St. Cloud, Paris, France
33(1) 771 03 08

Contact Person: Not supplied.

Content: This database contains information relating to market activity for a range of chemical and petrochemical products and liquified petroleum gas in Europe and the United States. Data are available in numeric time series or as textual reports with a paragraph commenting on the market activity. The file is comprised of three time series databases—GCG, ELPG, USLPG—and two additional reports listing the operating status of all ethylene crackers in Europe.

Format: Data/textual.

Time Span: History for all three time series is available from 1981.

Update Frequency: GCG, weekly; ELPG, two times/wk.; USLPG, daily; the crackers information, monthly.

Languages: English.

Vendor: I. P. Sharp Associates
2 First Canadian Place, Suite 1900
Toronto, ONT M5X 1E3, Canada
(416) 364-5361, (800) 387-1588; Telex: 0622259

Availability: Unrestricted.

Price Structure: Not supplied.

Downloading Restrictions: Not supplied.

Documentation: I. P. Sharp Energy Reference Guide.

Equivalent Printed Product: None.

Information supplied by vendor.

258. **OILS Database**

Producer: Petroleum Information
P.O. Box 158
Claremont, WA 6010, Australia
International +61 9 383 3477

Contact Person: Cathy Bagster.

Content: This database contains the latest progress report on every currently drilling oil exploration well in Australia, New Zealand, and Papua New Guinea, with an equity listing. Full historical and tenement data are available in the RIGS database (see entry 259), available from the same producer.

Format: Data/textual.

Time Span: Latest information only (historical data in the associated RIGS database).

Update Frequency: Daily.

Languages: English.

Vendor: ACI Computer Services
P.O. Box 42
Clayton, Victoria 3168, Australia
(03) 543 6166

8 — DATABASES

Availability: Unrestricted; telephone dial-up to Ausinet network in Australia.

Price Structure: Connect-time charge at $ A 90/hour; offline prints at $ A 1/document.

Downloading Restrictions: Not supplied.

Documentation: Data information sheet available from producer.

Equivalent Printed Product: Who's Drilling.

Information supplied by producer.

259. **RIGS Database**

Producer: Petroleum Information
P.O. Box 158
Claremont WA 6010, Australia
International +61 9 383 3477

Contact Person: Cathy Bagster.

Content: This database contains factual information (rather than references) on oil exploration in Australia and covers published and unpublished reports on every oil or gas well drilled in Australia. Current tenements with details of holders and past exploration within that tenement are listed in the file. A list of companies involved in exploration as equity partners, operators, service companies, and contractors with details of executive contacts, directors, company profile, shareholders, and company equity in tenements is also included. The OILS Database (approximately forty documents) is included at the front of the RIGS Database.

Format: Data/textual.

Time Span: Complete historical coverage of wells; seismic from 1980; all current companies, tenements, rigs, and basins.

Update Frequency: Weekly.

Languages: English.

Vendor: ACI Computer Services
P.O. Box 42
Clayton, Victoria 3168, Australia
(03) 543 6166

Availability: Unrestricted; telephone dial-up to Ausinet network in Australia.

Price Structure: Connect-time charge at $ A 90/hour; offline prints at $ A 1/document.

Downloading Restrictions: Not supplied.

Documentation: Data information sheet available from producer.

Equivalent Printed Product: Who's Drilling.

Information supplied by producer.

Full Text

260. **INFOIL 2**

   Producer: Norwegian Petroleum Directorate (NPD)
   Lagardsveien 80, P.O. Box 600
   N-4001 Stavanger, Norway
   (04) 53 31 60; Telex: 33100 NOPED

   Contact Person: Grete Schanche.

   Content: The INFOIL 2 database covers information on current British and Norwegian research and development projects in the offshore oil and gas subject area. This database contains descriptions of over 1,145 current Norwegian and British research projects in the areas of offshore oil, gas, and other petroleum-related industry activities. Abstracts of the projects and their intentions are included in the file.

   Format: Textual.

   Time Span: Current projects.

   Update Frequency: Continuous.

   Languages: English.

   Vendor: Norsk senter for informatikk A/S (NSI)
   Sidsel Aamodt Hoffen, P.O. Box 350 Blindern
   0314 Oslo 3, Norway
   (2) 45 25 08

   Availability: Unrestricted; request password from vendor.

   Price Structure: NOK 600/connect-hour.

   Downloading Restrictions: Not supplied.

   Documentation: Available from vendor.

8 – DATABASES

Equivalent Printed Product: None.

Information supplied by producer.

261. **Petroconsultants News Database**

   Producer: Computer Exploration Services, Burleigh House
   13 Newmarket Road
   Cambridge CB5 8EG, England
   (223) 315 933

   Contact Person: John Murray or Julian Young.

Content: Petroconsultants publishes a weekly telex service on northwest Europe and a weekly international exploration newsletter. Both of these publications are available online in a full-text database. Not an abstract or summary service, it contains the complete text of these two publications. *Weekly North West Europe* gives an up-to-date status report on each active exploration well in offshore United Kingdom, Netherlands, Norway, Ireland, Germany, Denmark, and France. The *International Oil Newsletter* concisely summarizes all activity on international (non-U.S.) exploration, production, and related upstream news.

Format: Textual.

Time Span: Weekly North West Europe, from 1982 to the present; *International Oil Newsletter*, from October 1984 to the present.

Update Frequency: Weekly.

Languages: English.

   Vendor: Petroconsultants, Inc.
   2 Houston Center, 909 Fannin, P-330
   Houston, TX 77010
   (713) 654-1368
       or
   Petroconsultants (UK) Ltd.
   30, St. James's Square
   London SW1Y 4JH, England
   (01) 930-5939

Availability: Subscription; request from producer/vendor.

Price Structure: Not supplied.

Downloading Restrictions: Not allowed except for local printing.

Documentation: User's guide.

Equivalent Printed Product: International Oil Newsletter; Weekly North West Europe.

Information supplied by producer.

Data Only

262. **Active Well Data On-line**

   Producer: Petroleum Information Corporation
   4100 E. Dry Creek Rd., P.O. Box 2612
   Denver, CO 80201
   (303) 740-7100

   Contact Person: Mike McCrory.

   Content: Petroleum Information's (PI) Active Well Data file contains information on the current status of wells which have received permits or are drilling. Online delivery of this constantly changing information enables users to easily track current well activity. Wells are entered into PI's Active Well Data file upon issuance of permit, and activity is carefully tracked through completion. Any additional activity following completion is also included. Operators are continuously contacted to verify current status. Current well data reports contain general well information listing location/identification data, test information including initial potential, production, drill stem and core tests, reported formation tops and corresponding depths, active drilling narrative, and more. Active Well Data is presently accessible online for select regions in the U.S. Regional coverage will increase as the system is developed.

   Format: Data.

   Time Span: Data on a particular well remain in the Active Well file for six months following completion or last reported date of activity. After completion, the well data may be retrieved from PI's Historical Well Data file.

   Update Frequency: File, daily; wells, weekly.

   Languages: English.

   Vendor: Producer.

   Availability: Request from producer.

   Price Structure: Request from producer.

   Downloading Restrictions: None.

   Documentation: Manuals, training folder, sample reports, code book, and quick reference guide.

94 / 8 – DATABASES

Equivalent Printed Product: Available from producer.

Information supplied by producer.

263. **ARGUS (Petroleum Argus)**

Producer: Petroleum Argus, Star House
104-108 Grafton Rd.
London NW5 1BD, England
(01) 485-8792; Telex: 21277 ARGUS G

Contact Person: Not supplied.

Content: The ARGUS database contains seven tables quoting the daily report, *Petroleum Argus.* It has bid-ask ranges for F.O.B. and C.I.F. spot prices for two regions — northwest Europe and the west Mediterranean. It covers gasoil, motor gasoline, naptha, kerosene-type jet fuel (JPI) and high- and low-sulfur heavy fuel oils. The RHINE database quotes Rhine barge and tankwagon prices for three regions (Rotterdam, Ruhr, and Basle) for gasoil, motor gasoline, and heavy fuel oil. The FREIGHT database quotes freight costs (netback calculations) for crude oil and petroleum products. The NYC database quotes New York City spot product prices for gasoil and motor gasoline. The GOVP database quotes daily barge prices for petroleum products, F.O.B., in Amsterdam, Rotterdam, and the Antwerp region. The CRUDE database has spot prices of Ras Tanura and Brent crude oil.

Format: Data.

Time Span: ARGUS daily from November 1979; RHINE semiweekly on Wednesdays and Fridays from January 1980; FREIGHT, NYC, GOVP weekly on Wednesdays from January 1979; CRUDE weekly on Wednesdays from January 1979 through September 1983 (Brent from January 1980) and daily from October 1983; BARGES daily from December 1982.

Update Frequency: RHINE, two times/wk. on the reporting days; ARGUS and BARGES, daily, before 16:00 UTC on the day following the reporting day; FREIGHT, NYC, CRUDE, and GOVP, weekly on the day following the reporting day.

Languages: English.

Vendor: I. P. Sharp Associates
2 First Canadian Place, Suite 1900
Toronto, ONT M5X 1E3, Canada
(416) 364-5361, (800) 387-1588; Telex: 0622259

Availability: Subscription; available from producer.

Price Structure: Not supplied.

Downloading Restrictions: Not supplied.

Documentation: I. P. Sharp Energy Reference Guide.

Equivalent Printed Product: Not supplied.

Information supplied by vendor.

264. **Banque de données du sous-sol français**

Producer: Bureau de Recherches Géologiques et Minières, Departement
Documentation et Information Géologiques
B.P. 6009
45060 ORLEANS CEDEX, France
(38) 64 38 09

Contact Person: J. P. Lepretre.

Content: This database contains references to factual data concerning French subsoil exploration contained in hydrogeological exploration, mineral and oil-gas exploration, and quarries. Entries include location, objective, type of data, borehole depth, and characteristics.

Format: Data.

Time Span: Not supplied.

Update Frequency: Two times/yr.

Languages: French.

Vendor: Questel, Inc. (beginning March 1986)
1625 I Street, NW, Suite 719
Washington, DC 20006-3001
(800) 424-9600
or (in Europe)
Telesystemes Questel
83,85 boulevard V. Ouriol
75013 Paris, France
33(1) 45 82 64 64

Availability: Request from producer.

Price Structure: Not supplied.

Downloading Restrictions: By special arrangement.

Documentation: User's guide (in French).

Equivalent Printed Product: No regular publication.

Information supplied by producer.

265. **BETC Crude Oil Analysis Data Bank**

Producer: National Institute for Petroleum and Energy Research
P.O. Box 2128
Bartlesville, OK 74005
(918) 336-2400

Contact Person: Paul W. Woodward.

Content: The data bank contains approximately nine thousand analyses. Although a conscious effort is made to cover all crude oils discovered in this country and representative crude oils from foreign countries, this goal is not completely attained. A crude oil analysis contains the general properties — gravity, sulfur content, nitrogen content, viscosity, color, and pour point, as well as location description of the source of the oil. The oil is also distilled up to an equivalent temperature of 42°C (45°F) intervals. Properties of these fractions are obtained where appropriate, such as gravity, refractive index, viscosity, and cloud point. Calculations can be made to obtain values for asphalt content, straight-run gasoline and its hydrocarbon composition, and many other desired values of properties.

Format: Data.

Time Span: Not supplied.

Update Frequency: Not supplied.

Languages: English.

Vendor: Producer.

Availability: Unrestricted; public access.

Price Structure: No connect-time charges to database.

Downloading Restrictions: Not supplied.

Documentation: Available from producer.

Equivalent Printed Product: BETC RI 78/14.

Information supplied by producer.

266. CREW (Seismic Crew Count)

Producer: Society of Exploration Geophysicists
P.O. Box 702740
Tulsa, OK 74170-2740
(918) 493-3516

Contact Person: Not supplied.

Content: The Seismic Crew Count database contains the monthly number of seismic land crews and marine vessels searching for oil and gas in the United States and U.S. waters, including Alaska. These figures are compiled from data reported by oil companies operating company-owned seismic exploration crews and vessels, and by contract seismic exploration companies. There are twelve time series in the database.

Format: Data.

Time Span: From May 1974 to the present.

Update Frequency: Monthly.

Languages: English.

Vendor: I. P. Sharp Associates
2 First Canadian Place, Suite 1900
Toronto, ONT M5X 1E3, Canada
(416) 364-5361, (800) 387-1588; Telex: 0622259

Availability: Unrestricted.

Price Structure: Not supplied.

Downloading Restrictions: Not supplied.

Documentation: I. P. Sharp Energy Reference Guide.

Equivalent Printed Product: Not supplied.

Information supplied by vendor.

267. DRI Natural Gas

Producer: Data Resources, Inc.
1750 K Street, NW, 9th Floor
Washington, DC 20006
(202) 862-3700

Contact Person: Diane H. Flasar.

Content: The DRI Natural Gas data bank contains financial and operating statistics for major U.S. interstate natural gas pipeline companies. Data reported on Federal Energy Regulatory Commission Forms 2, 11, and 15 are available from 1978 for the annual figures and from January 1982 for the monthly information. Data include assets and other debits; liabilities and other credits; natural gas sales, revenues, and reserves; gas delivered and received; operation and maintenance expenses; and number of customers.

Format: Data.

Time Span: Annual figures from 1978 to the present; monthly figures from 1982 to the present.

Update Frequency: In general series are updated as soon as new data are published by the source organization, usually within twenty-four hours of receipt of the data.

Languages: English.

Vendor: Producer.

Availability: Restricted; contact producer.

Price Structure: Data may be accessed through two pricing plans: annual subscription fee and Information Plus. Under the latter alternative, data are surcharged on a per-access basis, without the requirement of a fixed fee. Details on these price structures are available from Elliot Roseman in the DRI office in Lexington, Massachusetts, (617) 863-5100.

Downloading Restrictions: Selected sets may be downloaded to a personal computer for local processing.

Documentation: On-line and printed documentation is available for each of the DRI databases. Documentation provides short and long series names, frequency, time span, source, and other pertinent information to facilitate effective database use.

Equivalent Printed Product: Not supplied.

Information supplied by producer.

268. **DRI Oil and Gas Drilling**

Producer: Data Resources, Inc.
1750 K Street, NW, 9th Floor
Washington, DC 20006
(202) 862-3700

Contact Person: Diane H. Flasar.

Content: The DRI Oil and Gas Drilling data bank contains sixty-seven hundred time series which describe various activities of the oil and gas drilling industry. Data include total petroleum industry investments; number and footage of exploratory and development oil, gas, and dry wells drilled, and associated cost; number and footage of wildcat wells drilled; rotary rigs running and seismic crew counts; number of stripper wells, stripper well reserves, and stripper well production; proven reserves, reserve adjustments, and production of crude oil and natural gas; drilling and equipping well cost indices; and drilling intentions by state. Coverage includes state and substate regions and some international data. Annual, quarterly, monthly, and weekly figures are presented.

Format: Data.

Time Span: Varies.

Update Frequency: In general series are updated as soon as new data are published by the source organization, usually within twenty-four hours of receipt of the data.

Languages: English.

Vendor: Producer.

Availability: Restricted; contact producer.

Price Structure: Annual subscription fee. Details are available from Elliot Roseman in the DRI office in Lexington, Massachusetts, (617) 863-5100.

Downloading Restrictions: Selected sets may be downloaded to a personal computer for local processing.

Documentation: On-line and printed documentation is available for each of the DRI databases. Documentation provides short and long series names, frequency, time span, source, and other pertinent information to facilitate effective database use.

Equivalent Printed Product: Not supplied.

Information supplied by producer.

269. **DRI Oil Company**

Producer: Data Resources, Inc.
1750 K Street, NW, 9th Floor
Washington, DC 20006
(202) 862-3700

Contact Person: Diane H. Flasar.

Content: The DRI Oil Company data bank contains sixty-five hundred historical series which describe various activities of major oil and gas exploration companies. Data for each company include developed and undeveloped acreage; number of net wells, including gas, oil, dry, exploratory, and development wells; reserve data for crude oil and natural gas including year-end reserves, year-end developed reserves, additions, improved recovery, revisions, sale of in-place reserves and production; expenditures, including property acquisition, development, production, exploration, and windfall profit tax payments; average price for crude oil and natural gas; and runs to still. Coverage includes U.S. domestic activity by company in addition to international activity for selected companies. Information included in the database is taken from the annual reports of each company and company submissions of SEC Form 10-K to the Securities and Exchange Commission.

Format: Data.

Time Span: Varies.

Update Frequency: Not supplied.

Languages: English.

Vendor: Producer.

Availability: Restricted; contact producer.

Price Structure: Annual subscription fee. Details are available from Elliot Roseman in the DRI office in Lexington, Massachusetts, (617) 863-5100.

Downloading Restrictions: Selected sets may be downloaded to a personal computer for local processing.

Documentation: On-line and printed documentation is available for each of the DRI databases. Documentation provides short and long series names, frequency, time span, source, and other pertinent information to facilitate effective database use.

Equivalent Printed Product: Not supplied.

Information supplied by producer.

---

270. **DRI U.S. Energy**

 Producer: Data Resources, Inc.
 1750 K Street, NW, 9th Floor
 Washington, DC 20006
 (202) 862-3700

 Contact Person: Diane H. Flasar.

Content: The DRI U.S. Energy data banks provide state and national level data which describe the U.S. energy sector from resources and primary production through transformation into energy products to final consumption by sector. These statistics are supplemented by the State Energy Data System, which contains energy consumption data by fuel, end-market sector, and state as compiled by the U.S. Department of Energy. The Energy data banks provide an extensive collection of crude oil, coal, motor gasoline, naptha, nuclear power, and electricity statistics detailing exploration and production; consumption, demand, and deliveries; imports and exports; reserves, stocks, and inventories; refinery, gas plant, and utility operations; and retail and wholesale prices and sales. Annual, quarterly, monthly, and weekly statistics are presented in forty-one thousand series and cover the United States, census region, Petroleum Administration for Defense District, and state data.

Format: Data.

Time Span: Varies.

Update Frequency: In general series are updated as soon as new data are published by the source organization, usually within twenty-four hours of receipt of the data.

Languages: English.

Vendor: Producer.

Availability: Restricted; contact producer.

Price Structure: Data may be accessed through two pricing plans: annual subscription fee and Information Plus. Under the latter alternative, data are surcharged on a per-access basis, without the requirement of a fixed fee. Details on these price structures are available from Elliot Roseman in the DRI office in Lexington, Massachusetts, (617) 863-5100.

Downloading Restrictions: Selected sets may be downloaded to a personal computer for local processing.

Documentation: On-line and printed documentation is available for each of the DRI databases. Documentation provides short and long series names, frequency, time span, source, and other pertinent information to facilitate effective database use.

Equivalent Printed Product: Not supplied.

Information supplied by producer.

271. **Drilling Analysis Data On-line**

Producer: Petroleum Information Corporation
4100 E. Dry Creek Rd., P.O. Box 2612
Denver, CO 80201
(303) 740-7100

Contact Person: Mike McCrory.

Content: PI's Drilling Analysis Data file is a uniquely flexible database that allows statistical analyses of activity from 1970 to the present. Presented in a variety of established formats, some of the available data include well location, identification, type and class, success ratios, depth ranges, producing and total depth formation, competitor activity, and drilling and completion cost estimates. This nationwide database provides vital information for evaluating and analyzing prospects and opportunities by allowing the user to develop an historical perspective or isolate a slice in time. The file contains permits, well starts, completions, and suspensions. Drilling Analysis Data reports provide detailed statistics including drilled footage, cost per foot, and number of pays for wells identified by location and type. This online system also calculates cost of initial potential BOE/day, success ratios by depth, initial potential equivalent/foot, and more.

Format: Data.

Time Span: From 1970 to the present.

Update Frequency: Weekly.

Languages: English.

Vendor: Producer.

Availability: Request from producer.

Price Structure: Request from producer.

Downloading Restrictions: None.

Documentation: Manuals, training folder, sample reports, code book, and quick reference guide.

Equivalent Printed Product: Not supplied.

Information supplied by producer.

272. **Dwight's On-line System**

Producer: Dwight's Energydata, Inc.
1633 Firman Drive
Richardson, TX 75081
(214) 783-8002

Contact Person: Fred Smith.

Content: Dwight's On-line System is a computerized information system that provides direct access to a database of historical oil and gas production statistics. Production data for U.S. domestic properties, federal offshore areas, and Canada are available from the system. The database is updated with production statistics that have been gathered from government regulatory agencies. In addition to production statistics, the system provides other information to the user such as pressure test data and individual well or lease descriptions.

Format: Data.

Time Span: From 1934 to the present.

Update Frequency: Monthly.

Languages: English.

Vendor: General Electric Information Services Company
401 North Washington Street
Rockville, MD 20850
or
Energy Enterprises
1580 Lincoln, Suite 1000
Denver, CO 80203

Availability: Subscription available from producer.

Price Structure: Varies depending on type of information user wishes to access.

Downloading Restrictions: None.

Documentation: User's guide.

Equivalent Printed Product: Available by geographic regions in printed form, microfiche, and magnetic tapes.

Information supplied by producer.

104 / 8 — DATABASES

273. **EDPRICE (Lundberg Survey Energy Detente International Price and Tax Series)**

    Producer: Lundberg Survey, Inc.
    P.O. Box 3996
    North Hollywood, CA 91609
    (818) 768-5111

    Contact Person: Not supplied.

    Content: EDPRICE contains average prices and taxes for twenty-four petroleum and energy-related products in sixty-nine countries, as collected by Lundberg Survey Inc. Lundberg also provides extensive footnotes to aid in interpreting the data.

    Format: Data.

    Time Span: Monthly from October 1980.

    Update Frequency: Lundberg releases data for the Eastern Hemisphere on approximately the tenth of each month. They release data for the Western Hemisphere on the twenty-fifth of each month.

    Languages: English.

    Vendor: I. P. Sharp Associates
    2 First Canadian Place, Suite 1900
    Toronto, ONT M5X 1E3, Canada
    (416) 364-5361, (800) 387-1588; Telex: 0622259

    Availability: Subscription; available from producer.

    Price Structure: Not supplied.

    Downloading Restrictions: Not supplied.

    Documentation: I. P. Sharp Energy Reference Guide.

    Equivalent Printed Product: Energy Detente.

    Information supplied by vendor.

274. **EMIS (Electronic Markets and Information Systems Inc.)**

    Producer: McGraw-Hill Publications Company
    1221 Avenue of the Americas
    New York, NY 10020
    (212) 512-6143; Telex: 225977 EMIS UR

    Contact Person: Keith A. Haurie.

Content: This international information and trading service is a useful source of information for producers, marketers, and purchasers of crude oil, petroleum products, chemicals, fertilizers, and metals. It contains the most comprehensive energy information from all major industry sources on 350 related commodities. Also included are real-time industry and world news, up-to-date spot prices, live futures prices and commentary, financial and currency information, and shipping and storage data. Graphics, communications, and analytical facilities are provided.

Format: Data.

Time Span: From October 1982 to the present.

Update Frequency: Varies, depending on information (continual updates of futures prices, weekly updates of other data).

Languages: English.

Vendor: Bonneville Telecommunications
19 West South Temple
Salt Lake City, UT 84101
(801) 532-3400
  or
I. P. Sharp Associates
2 First Canadian Place, Suite 1900
Toronto, ONT M5X 1E3, Canada
(416) 364-5361, (800) 387-1588; Telex: 0622259

Availability: Request from producer.

Price Structure: Varies depending on what information user wishes to access. For details, contact producer.

Downloading Restrictions: Downloading capabilities are available. Transmission to nonsubscribers is discouraged.

Documentation: System menu, product list, and user's guide.

Equivalent Printed Product: Not supplied.

Information supplied by producer.

275. **Energy Historical Database**

Producer: Chase Econometrics
150 Monument Road
Bala Cynwyd, PA 19004-1780
(215) 667-6000

Contact Person: Anne McKeough.

Content: This database provides comprehensive coverage of major energy commodities, permitting in-depth analysis of the national and regional energy markets. Coverage ranges from general measures of supply and demand to energy-specific categories such as reserves, drilling, and refinery operations. Particular emphasis is on the petroleum industry. Extensive state-level statistics on energy consumption by major economic sector and by principal energy type are included. Selected international data are available. Timely updates are included of key weekly series such as crude oil spot prices.

Format: Data.

Time Span: Majority of data begin in 1960.

Update Frequency: According to data frequency.

Languages: English.

Vendor: Producer.

Availability: Not supplied.

Price Structure: Not supplied.

Downloading Restrictions: Request from producer.

Documentation: Available from producer.

Equivalent Printed Product: Not supplied.

Information supplied by producer.

276. **Historical Well Data On-line**

Producer: Petroleum Information Corporation
4100 E. Dry Creek Rd., P.O. Box 2612
Denver, CO 80201
(303) 740-7100

Contact Person: Mike McCrory.

Content: PI's computerized well histories provide a basis for studies that support future petroleum exploration, development, and production projects. All completed oil, gas, dry, service, and re-entry wells are included. Historical Well Data reports contain hundreds of detailed well information elements including location, classification, reported formation tops and corresponding depths, testing information covering initial potential, production, drill stem, shows, core tests, and more. Detailed engineering data on casing, cement, temperatures, and pressures are also available. Historical Well Data is useful for evaluating past drilling activity from permit to completion in any specific area. PI Historical Well Data On-line is available for wells in select regions.

Format: Data.

Time Span: 1.7 million well histories available.

Update Frequency: Monthly.

Languages: English.

Vendor: Producer.

Availability: Request from producer.

Price Structure: Request from producer.

Downloading Restrictions: None.

Documentation: Manuals, training folder, sample reports, code book, and quick reference guide.

Equivalent Printed Product: Available from producer.

Information supplied by producer.

277. **HUGHES (Hughes Rotary Drilling Rig Reports)**

Producer: International Association of Drilling Contractors
P.O. Box 4287
Houston, TX 77210
(713) 578-7171

Contact Person: Not supplied.

Content: This database provides counts of exploratory drilling rigs worldwide. There are two databases: HUGHESN and HUGHESI. HUGHESN has weekly data by state and regions for the United States as well as regional data for eastern and western Canada. HUGHESI contains monthly land and offshore statistics for ninety-eight countries in Africa, Europe, the Far East, the Middle East, South America, Central America, and North America. The data are reported by the Hughes Tool Company to the International Association of Drilling Contractors. There are 96 time series in HUGHESN and 294 time series in HUGHESI.

Format: Data.

Time Span: HUGHESN 1973 to the present; HUGHESI 1981 to the present.

Update Frequency: HUGHESN, weekly; HUGHESI, monthly.

Languages: English.

Vendor: I. P. Sharp Associates
2 First Canadian Place, Suite 1900
Toronto, ONT M5X 1E3, Canada
(416) 364-5361, (800) 387-1588; Telex: 0622259

Availability: Unrestricted.

Price Structure: Not supplied.

Downloading Restrictions: Not supplied.

Documentation: I. P. Sharp Energy Reference Guide.

Equivalent Printed Product: Hughes Rotary Rig Reports.

Information supplied by vendor.

278. **IMPORTS (Imports of Crude Oil and Petroleum Products)**

Producer: American Petroleum Institute
1220 L Street, NW
Washington, DC 20005
(202) 682-8000

Contact Person: Not supplied.

Content: IMPORTS gives information about every shipment into the United States of crude oil, residual, unfinished, and finished petroleum products. The database contains thirteen hundred time series. Volumes are given in barrels. Up to fifteen facts are associated with each shipment, including country of origin, port of entry, destination, importing and receiving companies, quantity in barrels, commodity, API gravity, sulfur content, and viscosity.

Format: Data.

Time Span: From 1977 to the present.

Update Frequency: Monthly.

Languages: English.

Vendor: I. P. Sharp Associates
2 First Canadian Place, Suite 1900
Toronto, ONT M5X 1E3, Canada
(416) 364-5361, (800) 387-1588; Telex: 0622259

Availability: Only subscribers to API's *Imported Crude Oil and Petroleum Products Report* may use this database.

Price Structure: Not supplied.

Downloading Restrictions: Not supplied.

Documentation: I. P. Sharp Energy Reference Guide.

Equivalent Printed Product: Imported Crude Oil and Petroleum Products Report.

Information supplied by vendor.

279. **KENTV (Kent Canadian Retail Gasoline Volume)**

Producer: Kent Marketing Services Ltd.
227 Colborne St.
London, ONT N6B 2S4, Canada

Contact Person: Not supplied.

Content: This database supplies gasoline volume and price information collected from approximately ninety urban market areas across Canada. Data from every retail outlet in each market is recorded, representing approximately 75 percent of the total gasoline volume sold in Canada. Time series data include volume and price for each of regular leaded, regular unleaded, premium (leaded and unleaded combined), and diesel gasoline. There are also over thirty static facts for each outlet, including station number, address, brand, outlet type, and total number of each type of pump.

Format: Data.

Time Span: From December 1980 to the present.

Update Frequency: Bimonthly for the twenty major markets; quarterly for the rest.

Languages: English.

Vendor: I. P. Sharp Associates
2 First Canadian Place, Suite 1900
Toronto, ONT M5X 1E3, Canada
(416) 364-5361, (800) 387-1588; Telex: 0622259

Availability: Subscription; available from producer.

Price Structure: Not supplied.

Downloading Restrictions: Not supplied.

Documentation: Energy Newsletter no. 9 (October 1983); online directory and online general description.

Equivalent Printed Product: None.

Information supplied by vendor.

280. **LOR (London Oil Reports)**

Producer: London Oil Reports
51 Harrowby St.
London W1, England
(01) 723 7997; Telex: 296557 LOR G

Contact Person: Not supplied.

Content: LOR provides spot prices and market activity in Europe and the United States for a range of crude oil types and petroleum products. In addition, extracts from the London Oil Reports' weekly newsletter analyze trends in the international oil industry. Netback values are available for crude oils in both markets. Four time series databases are available.

Format: Data.

Time Span: Varies with each of the four time series databases.

Update Frequency: Daily.

Languages: English.

Vendor: I. P. Sharp Associates
2 First Canadian Place, Suite 1900
Toronto, ONT M5X 1E3, Canada
(416) 364-5361, (800) 387-1588; Telex: 0622259

Availability: Unrestricted.

Price Structure: Not supplied.

Downloading Restrictions: Not supplied.

Documentation: I. P. Sharp Energy Reference Guide.

Equivalent Printed Product: Not supplied.

Information supplied by vendor.

281. **LPGAS (Liquefied Petroleum Gas)**

   Producer: American Petroleum Institute
   1220 L Street, NW
   Washington, DC 20005
   (202) 682-8000

   Contact Person: Not supplied.

   Content: This database contains inventories of liquefied petroleum gases (LPG) in all U.S. Petroleum Administration for Defense Districts. It includes inventories of stocks in plants, terminals, refineries, and underground facilities. Products included are ethane, butane-propane mix, propane, isopentane, ethane-propane mix, other stocks, isobutane, unfractionated stream, normal butane, other butanes, and totals. The database has approximately one hundred time series.

   Format: Data.

   Time Span: From January 1977 to the present.

   Update Frequency: Monthly.

   Languages: English.

   Vendor: I. P. Sharp Associates
   2 First Canadian Place, Suite 1900
   Toronto, ONT M5X 1E3, Canada
   (416) 364-5361, (800) 387-1588; Telex: 0622259

   Availability: Unrestricted.

   Price Structure: Not supplied.

   Downloading Restrictions: Not supplied.

   Documentation: I. P. Sharp Energy Reference Guide.

   Equivalent Printed Product: Not supplied.

   Information supplied by vendor.

282. **MER (Monthly Energy Review)**

   Producer: Energy Information Administration
   1000 Independence Avenue, SW
   Washington, DC 20585
   (202) 252-8800

   Contact Person: Not supplied.

Content: MER contains one hundred time series covering energy consumption by sector; consumption, production, sales, imports, and exports of natural gas; production, domestic consumption, imports, exports, and stocks of bituminous, lignite, and anthracite coal; statistics and domestic prices and percentages of crude oil purchased at the wellhead; F.O.B. cost of crude oil imports from selected countries; and landed cost of crude oil imports from selected countries.

Format: Data.

Time Span: From January 1977 to the present.

Update Frequency: Monthly.

Languages: English.

Vendor: I. P. Sharp Associates
2 First Canadian Place, Suite 1900
Toronto, ONT M5X 1E3, Canada
(416) 364-5361, (800) 387-1588; Telex: 0622259

Availability: Unrestricted.

Price Structure: Not supplied.

Downloading Restrictions: Not supplied.

Documentation: I. P. Sharp Energy Reference Guide.

Equivalent Printed Product: Monthly Energy Review.

Information supplied by vendor.

283. **Oil & Gas Journal Energy Database**

Producer: Oil & Gas Journal
1421 South Sheridan Road, P.O. Box 1260
Tulsa, OK 74101
(918) 835-3161

Contact Person: Glenda E. Smith.

Content: The Oil & Gas Journal Energy Database is an online statistical library of time series data on the petroleum industry. Data have been collected from the Energy Information Administration, the major trade associations, *Oil & Gas Journal*, and other private sources, and stored in one centralized location. Topics covered are drilling and exploration, production, reserves, refining, imports and exports, stocks (inventories), demand and consumption, natural gas, prices, transportation, offshore, and other energy. Over twenty thousand time series are currently included in the database.

Format: Data.

Time Span: For most time series at least ten years of historical data are included, for some as much as forty years of data has been included.

Update Frequency: Continuous.

Languages: English.

Vendor: General Electric Information Services Co. (GEISCO)
5050 Quorum Dr., Suite 500
Dallas, TX 75240
(214) 788-8214
Contact: John Wilder

Availability: There is no subscription fee to the database; however, a contractual agreement with GEISCO is required.

Price Structure: GEISCO requires a monthly minimum charge of $40.00. The database pricing is based upon a per series accessed basis. Weekly data are $0.75 per series and all other series are $3.00 each.

Downloading Restrictions: None.

Documentation: User's manual and directory of series in the database.

Equivalent Printed Product: None.

Information supplied by producer.

284. **Permit Data On-line**

Producer: Petroleum Information Corporation
4100 E. Dry Creek Rd., P.O. Box 2612
Denver, CO 80201
(303) 740-7100

Contact Person: Mike McCrory.

Content: Petroleum Information maintains a daily record of new drilling permits for all wells nationwide (excluding Alaska). Permit Data On-line routinely includes lease name, well number and classification, projected depth and formation, operator name, address and code, field name, state, county, control number and filer name, title, and phone number.

Format: Data.

Time Span: Data can be retrieved for the last ninety days.

114 / 8—DATABASES

Update Frequency: Daily. Data are available within 24-48 hours of release by regulatory agencies.

Languages: English.

Vendor: Producer.

Availability: Request from producer.

Price Structure: Request from producer.

Downloading Restrictions: None.

Documentation: Available from producer.

Equivalent Printed Product: Available from producer.

Information supplied by producer.

285. **Petroconsultants Exploration Database**

Producer: Computer Exploration Services
Burleigh House,
13 Newmarket Road
Cambridge CB5 8EG, England
(223) 351 933

Contact Person: John Murray or Julian Young.

Content: The Exploration Database contains information on foreign exploration and development wells, concessions, oil and gas fields, and production. Well data include classification, status, total depth, operator and working interests, coordinates, casing program, period of activity, and geological province. Coverage for well data is worldwide excluding U.S. and communist areas. Concession information describes the situation, type of rights, award date, relinquishment schedule, licensees, license area, farmout data, production split, signature bonus, and other commitments. Oil and gas field data describe status, situation, type, operating group, unitization, concession, coordinates, field area, wells drilled, and current well status. Coverage for oil and gas fields includes over eighty-two hundred fields and discoveries for ninety-nine countries worldwide, excluding the U.S. and Canada. Production data list oil, gas, and condensate production, reported by month, for fields and countries. Some countries such as Mexico do not have data for individual fields.

Format: Data.

Time Span: Wells, 1900 to the present; concessions, all valid in or awarded since 1976; oil and gas fields, all known fields and discoveries except for the Eastern Bloc, where only the significant fields/discoveries are recorded; production, 1977 to the present.

Update Frequency: Wells, continuous, but online service is monthly only; concessions, monthly; oil and gas fields, quarterly; production, monthly.

Languages: English.

Vendor: Petroconsultants Group, S.A.
8-10 rue Muzy
1211 Geneva, Switzerland

Availability: Subscription available from producer/vendor.

Price Structure: Not supplied.

Downloading Restrictions: Not allowed except for local printing.

Documentation: User's guide.

Equivalent Printed Product: Foreign Scouting Service; Country Acreage and Activity Statistics; and various others.

Information supplied by producer.

286. **Petroconsultants International Drilling**

Producer: Petroconsultants Group, S.A.
P.O. Box 228
1211 Geneva 6, Switzerland
368811; Telex: 27763 PETR CH

Contact Person: Jean-Pierre Javoques.

Content: Detailed information is provided on the exploratory activities of more than three hundred international oil and gas drilling companies in 115 countries. Data are reported in the following aggregates: type of well drilled, well depth, offshore by water depth, offshore by rig type, completion type (oil, gas, dry), location (onshore, offshore), and operating company. Coverage of the database is worldwide except for North America and other selected countries.

Format: Data.

Time Span: Varies.

Update Frequency: Not supplied.

Languages: Not supplied.

Vendor: Data Resources, Inc.
1750 K Street, NW, 9th Floor
Washington, DC 20006
(202) 862-3700
Contact: Diane H. Flasar

Availability: Restricted; contact vendor.

Price Structure: Annual subscription fee. Details are available from Elliot Roseman in the DRI office in Lexington, Massachusetts, (617) 863-5100.

Downloading Restrictions: Selected sets may be downloaded to a personal computer for local processing.

Documentation: Online and printed documentation is available for each of the databases available from DRI. Documentation provides short and long series names, frequency, time span, source, and other pertinent information to facilitate effective database use.

Equivalent Printed Product: None.

Information supplied by vendor.

287. **Petroflash (Petroflash! Crude and Product Reports)**

Producer: Petroflash! Inc.
P.O. Box 798
Lakewood, NJ 08701
(201) 367-1600

Contact Person: Not supplied.

Content: The Petroflash! database gives spot price assessments of key world export crudes and U.S. domestic grade crudes, both on a delivered-to-key market and F.O.B. loading facilities basis. Refined product values for important crudes in crucial downstream markets are detailed. In addition, official and contract prices for key crudes as well as crude oil inventory data are given.

Format: Data.

Time Span: The most current version of each report is available.

Update Frequency: Up to three times/day.

Languages: English.

Vendor: I. P. Sharp Associates
2 First Canadian Place, Suite 1900
Toronto, ONT M5X 1E3, Canada
(416) 364-5361, (800) 387-1588; Telex: 0622259

Availability: Subscription; available from producer.

Price Structure: Not supplied.

Downloading Restrictions: Not supplied.

Documentation: I. P. Sharp Energy Reference Guide.

Equivalent Printed Product: None.

Information supplied by vendor.

288. **PetroScan**

Producer: United Communications Group
Oil Price Information Service Division
4550 Montgomery Ave., Suite 700N
Bethesda, MD 20814
(301) 656-6666

Contact Person: Joann Giannola.

Content: PetroScan contains ten files covering U.S. wholesale prices and market information for No. 2 oil, regular, unleaded and premium unleaded gasolines, and liquefied petroleum and propane.

Format: Data.

Time Span: Current data in most files; historical statistics from 1982 in one file.

Update Frequency: Continuous, daily, or weekly updates depending on the file selected.

Languages: English.

Vendor: Producer.

Availability: Subscription.

Price Structure: Connect-time charge $2.10/min. for 1200 baud and $1.30/min. for 300 baud.

Downloading Restrictions: Not supplied.

118 / 8—DATABASES

Documentation: User's manual.

Equivalent Printed Product: Oil Price Information Service Newsletter.

Information supplied by producer.

289. **PIW (Petroleum Intelligence Weekly)**

Producer: Petroleum Intelligence Weekly
One Times Square Plaza
New York, NY 10036
(212) 575-1242

Contact Person: Diane E. Munro.

Content: This database contains four data series from *Petroleum Intelligence Weekly*. Crude Oil Production (COP) has forty-four monthly time series of volumes of crude oil produced in thousands of forty-two-gallon barrels for producers and major exporters, and totals of natural gas liquids for OPEC and the world. Key Crude Prices (KCP) has nine time series showing the broad trend of basic oil price indicators for three main types of crude in international trade. Price indicators are representative averages on an F.O.B. or port of loading basis. The Spot Product Series (SPP) has forty-one time series of representative monthly spot prices for the primary petroleum products in six key refining centers. The Official Crude Price (OCP) series has official crude prices for 114 countries.

Format: Data.

Time Span: Varies with database.

Update Frequency: COP production figures are usually available six to seven weeks after the end of the month. KCP and SPP prices are available the third or fourth week of the month. SPP revisions for Rotterdam and Italy are available the first or second week of the following month. OCP data are updated as soon as official price changes are confirmed by PIW.

Languages: English.

Vendor: I. P. Sharp Associates
2 First Canadian Place, Suite 1900
Toronto, ONT M5X 1E3, Canada
(416) 364-5361, (800) 387-1588; Telex: 0622259

Availability: Subscription; available from producer.

Price Structure: Not supplied.

Downloading Restrictions: Not supplied.

Documentation: I. P. Sharp Energy Reference Guide.

Equivalent Printed Product: Petroleum Intelligence Weekly.

Information supplied by vendor.

290. **Platt's Global Alert**

Producer: McGraw-Hill Publications Company
1221 Avenue of the Americas
New York, NY 10020
(212) 512-6571; Telex: 225977 EMIS UR

Contact Person: Robert S. Christie.

Content: This international petroleum price and news reporting service is available both online and via broadcast. Continuously updated petroleum prices and news, live futures prices and analysis, crude oil yields and netbacks, financial information, wholesale rack prices, latest spot transactions, and API statistics are included.

Format: Data.

Time Span: Current market data.

Update Frequency: Continuous.

Languages: English.

Vendor: I. P. Sharp Associates
2 First Canadian Place, Suite 1900
Toronto, ONT M5X 1E3, Canada
(416) 364-5361, (800) 387-1588; Telex: 0622259

Availability: Request from producer.

Price Structure: Varies depending on what information user wishes to access. For details, contact producer.

Downloading Restrictions: Downloading capabilities are available. Transmission to nonsubscribers is discouraged.

Documentation: User's guide and content directory.

Equivalent Printed Product: Not supplied.

Information supplied by producer.

120 / 8—DATABASES

291. **Platt's Oil Prices Databank**

   Producer: Data Resources, Inc.
   1750 K Street, NW, 9th Floor
   Washington, DC 20006
   (202) 862-3700

   Contact Person: Diane H. Flasar.

   Content: The Platt's Oil Prices Databank details domestic and international crude oil and petroleum product spot prices. Official and spot prices are available by type of foreign crude; product prices are available by market locations, disaggregated by sulfur content. International prices encompass European bulk cargo rates, tanker, and barge quotations. Domestic and European chemical prices for the U.S. Gulf Coast markets are offered. Fuels include crude oil, motor gasoline, heating oil, distillate fuels, and residual fuels. The data bank covers both U.S. and international petroleum markets in three thousand series. Monthly, weekly, and daily prices are included. Information is supplied by McGraw-Hill Publications Company.

   Format: Data.

   Time Span: Varies.

   Update Frequency: In general series are updated as soon as new data are published by the source organization, usually within twenty-four hours of receipt of the data.

   Languages: English.

   Vendor: Producer.

   Availability: Restricted; contact producer.

   Price Structure: Data may be accessed through two pricing plans: annual subscription fee and Information Plus. Under the latter alternative, data are surcharged on a per-access basis, without the requirement of a fixed fee. Details on these price structures are available from Elliot Roseman in the DRI office in Lexington, Massachusetts, (617) 863-5100.

   Downloading Restrictions: Selected sets may be downloaded to a personal computer for local processing.

   Documentation: Online and printed documentation is available for each of the DRI databases. Documentation provides short and long series names, frequency, time span, source, and other pertinent information to facilitate effective database use.

Equivalent Printed Product: Not supplied.

Information supplied by producer.

292. **PMPRICE (Petroleum Marketing Monthly)**

   Producer: Energy Information Administration
   U.S. Department of Energy
   Washington, DC 20585
   (202) 252-8800

   Contact Person: Not supplied.

   Content: Includes seven databases containing wholesale and retail sales prices for gasoline and petroleum products, plus volume of first sales of petroleum products for consumption. Most tables break down the data to the state level.

   Format: Data.

   Time Span: From January 1983 to the present.

   Update Frequency: Monthly.

   Languages: English.

   Vendor: I. P. Sharp Associates
   2 First Canadian Place, Suite 1900
   Toronto, ONT M5X 1E3, Canada
   (416) 364-5361, (800) 387-1588; Telex: 0622259

   Availability: Unrestricted.

   Price Structure: Not supplied.

   Downloading Restrictions: Not supplied.

   Documentation: I. P. Sharp Energy Reference Guide.

   Equivalent Printed Product: Petroleum Marketing Monthly.

   Information supplied by vendor.

293. **Production Data On-line**

   Producer: Petroleum Information Corporation
   4100 E. Dry Creek Rd., P.O. Box 2612
   Denver, CO 80201
   (303) 740-7100

   Contact Person: Mike McCrory.

Content: Petroleum Information's Production Data file contains current and historical monthly production volumes and crude cumulatives from inception, test data for crude and gas (where available), and field and reservoir summary data. PI offers the flexibility of combining crude and gas data in one report. This eliminates the need to run different reports for each product. PI production information is collected from state and federal agencies and edited to final format. Production reports provide comprehensive information on the field/reservoir, lease/well, and operator of interest. The file includes cumulative production totals and test data.

Format: Data.

Time Span: Current and historical production figures.

Update Frequency: Monthly.

Languages: English.

Vendor: Producer.

Availability: Request from producer.

Price Structure: Request from producer.

Downloading Restrictions: None.

Documentation: Manuals, training folder, sample reports, code book, and quick reference guide.

Equivalent Printed Products: Available from producer.

Information supplied by producer.

294. **QOS (Quarterly Oil Statistics)**

Producer: Organization for Economic Co-operation and Development
2, rue Andre-Pascal
75775 Paris CEDEX 16, France

Contact Person: Not supplied.

Content: The QOS database contains about 57,500 time series covering the twenty-five member countries of the OECD. The information includes balances of production, trade, refinery intake and output, final consumption, stock levels and changes for crude oil, natural gas liquids, feedstocks, and nine product groups. The database has import and export figures by origin and destination for liquefied petroleum gases, naptha, and the main product groups. It includes information on natural gas supply and consumption. It also covers marine bunkers and deliveries to international civil aviation by product group.

Format: Data.

Time Span: From 1974 to 1982.

Update Frequency: This file is no longer being updated.

Languages: English.

Vendor: I. P. Sharp Associates
2 First Canadian Place, Suite 1900
Toronto, ONT M5X 1E3, Canada
(416) 364-5361, (800) 387-1588; Telex: 0622259

Availability: Unrestricted.

Price Structure: Not supplied.

Downloading Restrictions: Not supplied.

Documentation: I. P. Sharp Energy Reference Guide.

Equivalent Printed Product: Not supplied.

Information supplied by vendor.

295. **RETAIL (Lundberg Survey Retail Prices)**

Producer: Lundberg Survey, Inc.
P.O. Box 3996
North Hollywood, CA 91609
(818) 768-5111

Contact Person: Not supplied.

Content: RETAIL includes retail prices for four grades of gasoline, diesel, and gasohol for over seventeen thousand fuel outlets in approximately seventy-five U.S. markets. Approximately fifty static facts describe the physical attributes and services available for each fuel outlet.

Format: Data.

Time Span: From 1981 to the present.

Update Frequency: Two times/mo.

Languages: English.

124 / 8—DATABASES

Vendor: I. P. Sharp Associates
2 First Canadian Place, Suite 1900
Toronto, ONT M5X 1E3, Canada
(416) 364-5361, (800) 387-1588; Telex: 0622259

Availability: Subscription; available from producer.

Price Structure: Not supplied.

Downloading Restrictions: Not supplied.

Documentation: I. P. Sharp Energy Reference Guide.

Equivalent Printed Product: Not supplied.

Information supplied by vendor.

296. **SOM (Lundberg Survey Share of Market)**

Producer: Lundberg Survey, Inc.
P.O. Box 3996
North Hollywood, CA 91609
(818) 768-5111

Contact Person: Not supplied.

Content: SOM contains sales volumes and percentage share of market for U.S. brands of gasoline, as calculated by Lundberg Survey, Inc. SOM includes data for over six thousand brands, for all fifty states and the District of Columbia, providing a total of approximately ten thousand state-brand combinations. (See also entry 297 for SOMSUM).

Format: Data.

Time Span: From 1978 to the present.

Update Frequency: Approximately weekly or as new data are received by the producer.

Languages: English.

Vendor: I. P. Sharp Associates
2 First Canadian Place, Suite 1900
Toronto, ONT M5X 1E3, Canada
(416) 364-5361, (800) 387-1588; Telex: 0622259

Availability: Subscription; available from producer.

Price Structure: Not supplied.

Downloading Restrictions: Not supplied.

Documentation: I. P. Sharp Energy Reference Guide.

Equivalent Printed Product: Not supplied.

Information supplied by vendor.

297. **SOMSUM (Lundberg Survey Share of Market Summary)**

Producer: Lundberg Survey, Inc.
P.O. Box 3996
North Hollywood, CA 91609
(818) 768-5111

Contact Person: Not supplied.

Content: In addition to the twenty thousand time series in SOM (see entry 296), approximately one thousand time series are available in this summary database. SOMSUM contains the sales volumes for seventeen major brands plus a total for the major brands and a total for all brands. The summary sales volumes are available for all states in the United States plus the five Petroleum Administration for Defense Districts.

Format: Data.

Time Span: From 1978 to the present.

Update Frequency: Approximately weekly or as new data are received by the producer.

Languages: English.

Vendor: I. P. Sharp Associates
2 First Canadian Place, Suite 1900
Toronto, ONT M5X 1E3, Canada
(416) 364-5361, (800) 387-1588; Telex: 0622259

Availability: Subscription; available from producer.

Price Structure: Not supplied.

Downloading Restrictions: Not supplied.

Documentation: I. P. Sharp Energy Reference Guide.

Equivalent Printed Product: Not supplied.

Information supplied by vendor.

126 / 8 – DATABASES

298. **USDOE (United States Department of Energy)**

    Producer: Energy Information Administration
    U.S. Department of Energy
    Washington, DC 20585
    (202) 252-8800

    Contact Person: Not supplied.

    Content: USDOE contains information on stocks, production, imports of crude oil and petroleum products, and refinery operations in the United States. The database covers supply and disposition of crude oil and petroleum products; production of crude oil, lease condensate, and natural gases; refinery receipts, input, production, yield, and fuel consumption; imports and exports of crude oil and petroleum products; and production, stocks, and imports of heavy fuel oil by sulfur content. Most tables break down the data to the Petroleum Administration for Defense District or further.

    Format: Data.

    Time Span: From 1972 to the present.

    Update Frequency: Monthly.

    Languages: English.

    Vendor: I. P. Sharp Associates
    2 First Canadian Place, Suite 1900
    Toronto, ONT M5X 1E3, Canada
    (416) 364-5361, (800) 387-1588; Telex: 0622259

    Availability: Unrestricted.

    Price Structure: Not supplied.

    Downloading Restrictions: Not supplied.

    Documentation: I. P. Sharp Energy Reference Guide.

    Equivalent Printed Product: Not supplied.

    Information supplied by vendor.

299. **Weekly Statistical Bulletin**

    Producer: American Petroleum Institute
    1220 L Street, NW
    Washington, DC 20005
    (202) 682-8000

Contact Person: J. J. Tsikerdanos.

Content: This database contains U.S. data relating to refinery activity and principal inventories, crude oil and product imports, crude oil production, etc. Once a month, an estimated *Supply and Demand Situation Report* analyzing and commenting on the significance of trends reflected in the weekly data will be included in the *Bulletin.* The monthly report is published each month in the second or third *Weekly Bulletin* following the reporting month. A commentary will accompany the monthly statistical summary.

Format: Data.

Time Span: From January 1976 to the present.

Update Frequency: Weekly on Tuesday.

Languages: English.

Vendor: I. P. Sharp Associates (for database)
2 First Canadian Place, Suite 1900
Toronto, ONT M5X 1E3, Canada
(416) 364-5361, (800) 387-1588; Telex: 0622259

American Petroleum Institute (for equivalent printed product)
1220 L Street, NW
Washington, DC 20005
(202) 682-8000

Availability: Unrestricted; available from vendor.

Price Structure: Not supplied.

Downloading Restrictions: Not supplied.

Documentation: Available from producer and vendor.

Equivalent Printed Product: Weekly Statistical Bulletin and *Monthly Statistical Report.*

Information supplied by producer.

300. **Well Completions Database**

Producer: Oil & Gas Journal
1421 South Sheridan Road, P.O. Box 1260
Tulsa, OK 74101
(918) 835-3161

Contact Person: Glenda E. Smith.

128 / 8—DATABASES

Content: The Well Completions Database contains data on well completions from the American Petroleum Institute (API). Data are available back to 1970 on oil, gas, and service wells and dry holes reported by state, county, and depth interval. Breakouts of exploratory and development wells are provided. Data are provided on both the "as reported" and "as completed" reporting basis. Information is aggregated by reporting entity so no individual well data are available. Information may be obtained for both the number of wells drilled by type and the total footage.

Format: Data.

Time Span: From 1970 to the present.

Update Frequency: Monthly.

Languages: English.

Vendor: General Electric Information Services Co. (GEISCO)
   5050 Quorum Dr., Suite 500
   Dallas, TX 75240
   (214) 788-8214
   Contact: John Wilder

Availability: There is no subscription fee to the database; however, a contractual agreement with GEISCO is required.

Price Structure: GEISCO requires a monthly minimum charge of $40.00.

Downloading Restrictions: Not supplied.

Documentation: Documentation for this database is included in the user's manual for the Oil & Gas Energy Database (see entry 283).

Equivalent Printed Product: None.

Information supplied by producer.

301. **WHOLESALE (Lundberg Survey Wholesale Prices and Moves)**

   Producer: Lundberg Survey, Inc.
      P.O. Box 3996
      North Hollywood, CA 91609
      (818) 768-5111

   Contact Person: Not supplied.

   Content: This database contains wholesale prices of gasoline and distillate No. 2 for over two hundred regions in the United States. WHOLESALE includes all brands holding at least a 2 percent share of their respective U.S. markets. It contains forty thousand time series prices and their effective dates.

8 – DATABASES / 129

Format: Data.

Time Span: Both current and historical prices are available. The current price is the latest price reported to Lundberg Survey Inc. Historical prices are the final prices and effective dates for each Friday for the most recent twenty-four months.

Update Frequency: Continuous.

Languages: English.

Vendor: I. P. Sharp Associates
2 First Canadian Place, Suite 1900
Toronto, ONT M5X 1E3, Canada
(416) 364-5361, (800) 387-1588; Telex: 0622259

Availability: Subscription; available from producer.

Price Structure: Not supplied.

Downloading Restrictions: Not supplied.

Documentation: I. P. Sharp Energy Reference Guide.

Equivalent Printed Product: Not supplied.

Information supplied by vendor.

9 Periodicals

Petroleum-related periodicals publish information on current technology, research results, forecasts, historical perspectives, continuing education opportunities, people, and events. They may also include book reviews, calendars, product reviews, and statistics. Some periodicals publish all of this information in each issue; others publish special issues presenting forecasts, directories, products, statistics, etc. These special issues are easily identified by using reference books such as the *Special Issues Index*[1] or *Guide to Industry Special Issues*.[2]

This chapter contains a carefully selected list of petroleum-related periodicals. Each entry includes title, price, beginning date, frequency of publication, publisher information, and the International Standard Serial Number (ISSN). All information has been verified either using a 1986 issue of the publication or from *Ulrich's International Periodicals Directory*.[3] The list demonstrates the wide variety of titles available to the petroleum industry. It is not intended to be comprehensive, but to provide a good foundation for any petroleum periodical collection. Several basic reference books will help the reader expand or update the list found in this chapter; *Ulrich's International Periodicals Directory* and *The Standard International Periodicals Directory*[4] are just two examples.

---

[1]*Special Issues Index: Specialized Contents of Business, Industrial, and Consumer Journals* (Westport, Conn.: Greenwood Press, 1982). 309p. ISBN 0-313-23278-4.

[2]*Guide to Industry Special Issues and Indexes of Periodicals* (Washington, D.C.: Special Libraries Association, 1985). 160p. ISBN 0-87111-263-9.

[3]*Ulrich's International Periodicals Directory* (New York: R. R. Bowker, 1985). LC 32-16320. ISSN 0000-0175.

[4]*The Standard International Periodicals Directory* (New York: Oxbridge Communications, Inc., 1985). LC 64-7598. ISSN 0085-6630.

9 – PERIODICALS / 131

302. **AAPG Bulletin.** 1917- . monthly, twice in January. $70.00/yr. (U.S.); $95.00/yr. (foreign). ISSN 0149-1423.
    American Association of Petroleum Geologists
    1444 South Boulder Avenue
    Tulsa, OK 74119-3604

303. **AAPG Explorer.** 1979- . monthly. $15.00/yr. ISSN 0195-2986.
    American Association of Petroleum Geologists
    1444 South Boulder Avenue
    Tulsa, OK 74119-3604

304. **AICHE Journal.** 1955- . monthly. $40.00/yr. (members); $250.00/yr. (non-members). ISSN 0001-1541.
    American Institute of Chemical Engineers
    345 East 47th Street
    New York, NY 10017

305. **American Gas Association Monthly.** 1919- . 11 issues/yr. $30.00/yr. ISSN 0002-8584.
    American Gas Association
    1515 Wilson Blvd.
    Arlington, VA 22209

306. **American Journal of Science.** 1818- . 10 issues/yr. $40.00/yr. (individuals); $80.00/yr. (institutions). ISSN 0002-9599.
    Kline Geology Laboratory
    Yale University
    New Haven, CT 06511

307. **APEA Journal.** 1961- . 2 issues/yr. $80.00/yr. (Australian). ISSN W084-7534.
    Australian Petroleum Exploration Association
    G.P.O. Box 3974
    Sydney, NSW 2001, Australia

308. **BMR Journal of Australian Geology & Geophysics.** 1976- . quarterly. $27.00/yr. (Australian). ISSN 0312-9608.
    Australian Government Publishing Office
    G.P.O. Box 84
    Canberra, ACT 2601, Australia

309. **Bulletin of Canadian Petroleum Geology.** 1953- . quarterly. $8.00/yr. (Canadian). ISSN 0007-4802.
    Canadian Society of Petroleum Geologists
    505, 206 7th Avenue S.W.
    Calgary, ALB T2P 0W7, Canada

310. **Canadian Journal of Earth Sciences/Journal canadien des sciences de la terre.** 1964- . monthly. $78.00/yr. (Canadian; individuals); $164.00/yr. (Canadian; institutions). ISSN 0008-4077.
    Distribution, R-88 (Canadian Journal of Earth Sciences)
    National Research Council of Canada
    Ottawa, ONT K1A 0R6, Canada

132 / 9 – PERIODICALS

311. **Canadian Petroleum.** 1956- . monthly. $39.50/yr. ISSN 0008-4735.
 Southam Communications Ltd.
 1450 Don Mills Road
 Don Mills, ONT M3B 2X7, Canada

312. **Chemical & Engineering News.** 1923- . weekly, except last week of December. $35.00/yr. (nonmembers). ISSN 0009-2347.
 American Chemical Society
 1155 16th Street, NW
 Washington, DC 20036

313. **Chemical Engineering.** 1902- . biweekly. $24.50/yr. ISSN 0009-2460.
 McGraw-Hill, Inc., McGraw-Hill Building
 1221 Avenue of the Americas
 New York, NY 10020

314. **Chemical Engineering Progress.** 1947- . monthly. $35.00/yr. (nonmembers). ISSN 0360-7275.
 American Institute of Chemical Engineers
 345 East 47th Street
 New York, NY 10017

315. **China Oil.** 1984- . quarterly. $70.00/yr.
 Wen Wei Po Ltd., H.K.
 197 Wanchai Road, g/F., Hong Kong

316. **Clays and Clay Minerals.** 1968- . bimonthly. $96.00/yr. ISSN 0009-8604.
 The Clay Minerals Society
 P.O. Box 368
 Lawrence, KS 66044

317. **Computers & Chemical Engineering: An International Journal.** 1976- . bimonthly. $215.00/yr. ISSN 0098-1354.
 Pergamon Press, Inc.
 Maxwell House, Fairview Park
 Elmsford, NY 10523

318. **Computers & Geosciences.** 1975- . bimonthly. $235.00/yr. ISSN 0098-3004.
 Pergamon Press, Inc.
 Maxwell House, Fairview Park
 Elmsford, NY 10523

319. **Computers and Geotechnics: An International Journal.** 1985- . quarterly. $72.00/yr. ISSN 0266-352X.
 Elsevier Applied Science Publishers Ltd.
 Crown House, Linton Road
 Barking, Essex IG11 8JU, England

320. **Corrosion.** 1945- . monthly. $75.00/yr. ISSN 0010-9312.
National Association of Corrosion Engineers
1440 South Creek Drive
Houston, TX 77084

Drill Bit (ISSN 0012-6225). *See Southwest Oil World*

321. **Drilling Contractor.** 1944- . monthly. $35.00/yr. ISSN 0046-0702.
Drilling Contractor Publications, Inc.
P.O. Box 4287
Houston, TX 77210

322. **Drilling, The Wellsite Publication.** 1939- . monthly. $34.50/yr. ISSN 0012-6241.
Harcourt Brace Jovanovich
Energy Publications Division
P.O. Box 1589
Dallas, TX 75221-1589

323. **Earth and Planetary Science Letters.** 1965- . monthly. FL 1150/yr. ISSN 0012-821X.
Elsevier Scientific Publishing Co., Journals Department
P.O. Box 211
1000 AE Amsterdam, The Netherlands

324. **Economic Geology and The Bulletin of the Society of Economic Geologists.** 1906- . 8 issues/yr. $39.00/yr. ISSN 0361-0128.
Economic Geology Publishing Company
Kenneth F. Clark, business editor
P.O. Box 637
The University of Texas-El Paso
El Paso, TX 79968-0637

325. **The Energy Daily.** 1973- . daily, weekdays. $895.00/yr. ISSN 0364-5274.
King Publishing Group
915 15th Street, NW, Suite 400
Washington, DC 20005

326. **Energy Exploration & Exploitation.** 1982- . quarterly. $75.00/yr. ISSN 0144-5987.
Elsevier Science Publishing Co., Journal Information Center
52 Vanderbilt Avenue
New York, NY 10017

327. **Enhanced Recovery Week.** 1980- . weekly, except first week in January and second week in July. $345.00/yr. ISSN 0277-9137.
Pasha Publications
1401 Wilson Blvd., Suite 910
Arlington, VA 22209

134 / 9 – PERIODICALS

328. **EOS, Transactions, American Geophysical Union.** 1919- . weekly. $115.00/yr. (nonmembers). ISSN 0096-3941.
American Geophysical Union
2000 Florida Avenue, NW
Washington, DC 20009

329. **Fuel.** 1922- . monthly. $526.00/yr. ISSN 0016-2361.
Butterworth Scientific Ltd.
P.O. Box 63
Westbury House, Bury Street
Guildford, Surrey GU2 5BH, England

330. **Geochemistry International.** 1964- . (English translation of *Geokhimiya*. ISSN 0016-7525). 6 issues/yr. $488.00/yr. ISSN 0016-7029.
John Wiley & Sons
605 Third Avenue
New York, NY 10158

331. **Geochimica et cosmochimica acta.** 1950- . monthly. $340.00/yr. ISSN 0016-7037.
Pergamon Press, Inc.
Maxwell House, Fairview Park
Elmsford, NY 10523

Geokhimiya (ISSN 0016-7525). *See Geochemistry International*

332. **Geological Society of America Bulletin.** 1888- . monthly. $80.00/yr. ISSN 0016-7606.
Geological Society of America, Inc.
3300 Penrose Place
Boulder, CO 80301

333. **Geologiya i Geofizika.** 1960- . monthly. contact publisher for price. ISSN 0016-7886.
Akademiya Nauk S.S.S.R., Sibirskoe Otdelenie,
    Institut Geologii i Geofiziki
Novosibirsk, Akademgorodok, USSR

334. **Geologiya Nefti i Gaza.** 1957- . monthly. $18.50/yr. ISSN 0016-7894.
Izdatel'stvo Nedra Tretyakovskii proezd
1, Moscow K-12
Russian S.F.S.R., USSR

335. **Geology.** 1973- . monthly. $55.00/yr. (nonmembers). ISSN 0091-7613.
Geological Society of America, Inc.
3300 Penrose Place
Boulder, CO 80301

9 – PERIODICALS / 135

336. **Geophysical Prospecting.** 1953- . 8 issues/yr. $109.00/yr. ISSN 0016-8025.
Blackwell Scientific Publications Ltd.
P.O. Box 88
Oxford, England

337. **Geophysics.** 1936- . monthly. $45.00/yr. ISSN 0016-8033.
Society of Exploration Geophysicists
8801 South Yale
Tulsa, OK 74137

338. **Geoscience Canada.** 1974- . quarterly. $25.00/yr. (Canadian). ISSN 0315-0941.
GAC Publication Division, Business and Economic Service Ltd.
111 Peter Street, Suite 509
Toronto, ONT M5V 2H1, Canada

339. **Geotimes.** 1956- . monthly. $18.00/yr. ISSN 0016-8556.
American Geological Institute
4220 King Street
Alexandria, VA 22302

340. **Gulf Coast Oil World.** 1981- . monthly. $24.00/yr. ISSN 0884-7967.
Hart Publications, Inc.
1900 Grant Street, Suite 400, P.O. Box 1917
Denver, CO 80201-1917

341. **Hydrocarbon Processing.** 1922- . monthly. $250.00/yr. ISSN 0018-8190.
Gulf Publishing Company
P.O. Box 2608
Houston, TX 77001

342. **International Journal of Multiphase Flow.** 1974- . bimonthly. $250.00/yr. ISSN 0301-9322.
Pergamon Press, Inc., Journals Division
Maxwell House, Fairview Park
Elmsford, NY 10523

343. **Isotope Geoscience.** 1983- . 8 issues/yr. FL 450/yr. (institutions). ISSN 0167-6695.
Elsevier Scientific Publishing Co., Journals Department
P.O. Box 211
1000 AE Amsterdam, The Netherlands

344. **Japan Petroleum Weekly.** 1966- . weekly. $850.00/yr. ISSN 0386-6165.
Japan Petroleum and Energy Consultants, Ltd.
Nihon Sekiyu Konsarutanto K.K.
Box 1185, Tokyo Central
Tokyo 100-91, Japan

345. **The Journal of Canadian Petroleum Technology.** 1962- . 6 issues/yr. $30.00/yr. (Canadian). ISSN 0021-9487.
    Canadian Institute of Mining and Metallurgy
    400-1130 Sherbrooke Street W.
    Montreal, QUE H3A 2M8, Canada

346. **Journal of Colloid and Interface Science.** 1946- . monthly. $666.00/yr. (U.S.); $752.50/yr. (foreign). ISSN 0021-9797.
    Academic Press
    111 Fifth Avenue
    New York, NY 10003

347. **The Journal of Energy Resources Technology.** 1979- . quarterly. $36.00/yr. (members); $72.00/yr. (nonmembers). ISSN 0195-0738.
    American Society of Mechanical Engineers
    345 East 47th Street
    New York, NY 10017

348. **The Journal of Engineering for Industry.** 1970- . quarterly. $36.00/yr. (members); $72.00/yr. (nonmembers). ISSN 0022-0617.
    American Society of Mechanical Engineers
    345 East 47th Street
    New York, NY 10017

349. **The Journal of Engineering Materials and Technology.** 1972- . quarterly. $36.00/yr. (members); $72.00/yr. (nonmembers). ISSN 0094-4289.
    American Society of Mechanical Engineers
    345 East 47th Street
    New York, NY 10017

350. **Journal of Geochemical Exploration.** 1972- . 9 issues/yr. FL 400/yr. ISSN 0375-6742.
    Elsevier Scientific Publishing Co., Journals Department
    P.O. Box 211
    1000 AE Amsterdam, The Netherlands

351. **The Journal of Geology.** 1893- . bimonthly. $30.00/yr. (individuals); $45.00/yr. (institutions). ISSN 0022-1376.
    University of Chicago Press
    5801 Ellis Avenue
    Chicago, IL 60637

9 – PERIODICALS / 137

352. **Journal of Geophysical Research: Oceans and Atmospheres.** 1896- . monthly. $470.00/yr. (nonmembers). ISSN 0196-2256.
American Geophysical Union
2000 Florida Avenue, NW
Washington, DC 20009

353. **Journal of Geophysical Research: Solid Earth and Planets.** 1896- . monthly, twice in February. $630.00/yr. (nonmembers). ISSN 0148-0227.
American Geophysical Union
2000 Florida Avenue, NW
Washington, DC 20009

354. **Journal of Geophysics.** 1924- . 6 issues/yr. $174.00/yr. ISSN 0340-062X.
Springer-Verlag New York, Inc.
Service Center Secaucus
44 Hartz Way
Secaucus, NJ 07094

355. **Journal of Geotechnical Engineering.** 1956- . monthly. $77.00/yr. (nonmembers). ISSN 0733-9410.
American Society of Civil Engineers
345 East 47th Street
New York, NY 10017-2398

356. **Journal of Paleontology.** 1927- . bimonthly. $78.00/yr. (nonmembers). ISSN 0022-3360.
Society of Economic Paleontologists and Mineralogists
P.O. Box 4756
Tulsa, OK 74159-0756

357. **Journal of Petroleum Geology.** 1978- . quarterly. $130.00/yr. ISSN 0141-6421.
Scientific Press Ltd.
P.O. Box 21
Beaconsfield
Bucks HP9 1NS, England

358. **Journal of Petroleum Technology.** 1949- . monthly, twice in May. $24.00/yr. (nonmembers). ISSN 0149-2136.
Society of Petroleum Engineers
222 Palisades Creek Drive
Richardson, TX 75080

359. **Journal of Physical Chemistry.** 1896- . biweekly. $45.00/yr. (members); $268.00/yr. (nonmembers). ISSN 0022-3654.
American Chemical Society
1155 16th Street, NW
Washington, DC 20036

138 / 9 – PERIODICALS

360. **Journal of Sedimentary Petrology.** 1931- . bimonthly. $95.00/yr. (nonmembers). ISSN 0022-4472.
    Society of Economic Paleontologists and Mineralogists
    P.O. Box 4756
    Tulsa, OK 74159-0756

361. **Journal of Structural Engineering.** 1956- . monthly. $123.00/yr. ISSN 0733-9445.
    American Society of Civil Engineers
    345 East 47th Street
    New York, NY 10017-2398

362. **Journal of Structural Geology.** 1979- . quarterly. $160.00/yr. ISSN 0191-8141.
    Pergamon Press Ltd., Subscriptions Department
    Headington Hill Hall
    Oxford OX3 0BW, England

363. **The Journal of the Acoustical Society of America.** 1929- . monthly. $300.00/yr. ISSN 0001-4966.
    American Institute of Physics
    335 East 45th Street
    New York, NY 10017

364. **Journal of the American Oil Chemists' Society.** 1917- . monthly. $60.00/yr. ISSN 0003-021X.
    American Oil Chemists' Society
    508 S. Sixth Street
    Champaign, IL 61820

365. **Journal of the International Association for Mathematical Geology.** 1969- . 8 issues/yr. $225.00/yr. ISSN 0020-5958.
    Plenum Publishing Corporation
    233 Spring Street
    New York, NY 10013

366. **Journal of Transportation Engineering.** 1969- . bimonthly. $60.00/yr. (nonmembers). ISSN 0733-947X.
    American Society of Civil Engineers
    345 East 47th Street
    New York, NY 10017-2398

367. **The Log Analyst.** 1962- . bimonthly. $25.00/yr. ISSN 0024-581X.
    Society of Professional Well Log Analysts
    6001 Gulf Frwy., Suite C129
    Houston, TX 77023

368. **Marine and Petroleum Geology.** 1984- . quarterly. $245.00/yr. ISSN 0264-8172.
    Butterworth Scientific Ltd.
    P.O. Box 63
    Westbury House, Bury Street
    Guildford, Surrey GU2 5BH, England

369. **Marine Geology.** 1964- . 24 issues/yr. FL 1140/yr. ISSN 0025-3227.
Elsevier Scientific Publishing Co., Journals Department
P.O. Box 211
1000 AE Amsterdam, The Netherlands

370. **Marine Geotechnology.** 1975- . quarterly. $72.00/yr. ISSN 0360-8867.
Crane, Russak & Company, Inc.
3 East 44th Street
New York, NY 10017

371. **Marine Technology.** 1964- . quarterly. $40.00/yr. (nonmembers). ISSN 0025-3316.
Society of Naval Architects and Marine Engineers
One World Trade Center, Suite 1369
New York, NY 10048

372. **National Petroleum News.** 1909- . 13 issues/yr. $35.00/yr. ISSN 0027-9889.
Hunter Publishing Company
950 Lee Street
Des Plaines, IL 60016

373. **Nature.** 1869- . weekly, except last week in December. $173.00/yr. ISSN 0028-0836.
Macmillan Journals Ltd., Subscription Department
P.O. Box 1501
Neptune, NJ 07753

374. **New Scientist.** 1956- . weekly. $99.00/yr. ISSN 0028-6664.
New Science Publications, Holborn Publishing Group
Commonwealth House, 1-19 New Oxford Street
London WC1A 1NG, England

375. **Noroil.** 1973- . monthly. $130.00/yr. ISSN 0332-544X.
Noroil Publishing House Ltd. A/S
P.O. Box 480
Hillevagsvn 17
4001 Stavanger, Norway

376. **Ocean Industry.** 1966- . monthly. $15.00/yr. ISSN 0029-8026.
Gulf Publishing Company
P.O. Box 2608
Houston, TX 77001

377. **Offshore.** 1954- . monthly. $45.00/yr. ISSN 0030-0608.
PennWell Publishing Co.
1421 South Sheridan Road
Tulsa, OK 74101

140 / 9—PERIODICALS

378. **Offshore Engineer.** 1975- . monthly. £45.00/yr. (outside U.K.); £30.00 (U.K.). ISSN 0305-876X.
    Thomas Telford Ltd.
    P.O. Box 101, 26/34 Old Street
    London EC1P 1JH, England

379. **Oil & Gas Journal.** 1902- . weekly. $34.00/yr. ISSN 0030-1388.
    PennWell Publishing Co.
    1421 South Sheridan Road
    Tulsa, OK 74101

380. **Oil & Petrochemical Pollution.** 1982- . quarterly. $74.00/yr. ISSN 0143-7127.
    Elsevier Applied Science Publishers Ltd.
    Crown House, Linton Road
    Barking, Essex IG11 8JU, England

381. **The Oil Daily.** 1951- . daily, weekdays. $297.00/yr. ISSN 0030-1434.
    The Oil Daily Inc.
    1401 New York Avenue, NW, Suite 500
    Washington, DC 20005

382. **Oil Gas European Magazine.** 1975- . (English-language edition of *Erdoel-Erdgas Zeitschrift*). 2 issues/yr. DM 60,00/yr. ISSN 0342-5622.
    Urban-Verlag, Hamburg/Wien GmbH
    Neumann-Reichardt-Str. 34, P.O. Box 70 16 06
    D-2000 Hamburg 70, West Germany

383. **The Oilman.** 1973- . monthly. £35.00/yr. (surface mail); £65.00/yr. (airmail). ISSN 0143-6694.
    Maclean Hunter Ltd.
    76 Oxford Street
    London W1N 9FD, England

384. **Oilweek.** 1950- . weekly. $31.00/yr. (Canadian). ISSN 0030-1515.
    Maclean Hunter Ltd.
    200 - 1015 Centre St. N
    Calgary, ALB T2E 2P8, Canada

385. **Pace Synthetic Fuels Report.** 1964- . quarterly. contact publisher for price.
    The Pace Company Consultants & Engineers, Inc.
    P.O. Box 53473
    Houston, TX 77052

386. **Pacific Oil World.** 1908- . monthly. $15.00/yr. ISSN 0008-1329.
    Petroleum Publishers, Inc.
    222 South Brea Blvd., P.O. Box 129
    Brea, CA 92621

387. **Palaeogeography Palaeoclimatology Palaeoecology.** 1965- . 16 issues/yr. FL 800/yr. ISSN 0031-0182.
>Elsevier Scientific Publishing Co., Journals Department
P.O. Box 211
1000 AE Amsterdam, The Netherlands

388. **Palaeontology.** 1957- . quarterly. $32.00/yr. (membership); $18.00/yr. (student membership); $16.00/yr. (retired membership); $68.00/yr. (institutional membership). ISSN 0031-0239.
>The Palaeontological Association
Dr. A. T. Thomas, Department of Geological Sciences
University of Aston, Aston Triangle
Birmingham B4 7ET, England

389. **Petroleum Economist.** 1934- . monthly. $144.00/yr. ISSN 0306-395X.
>Petroleum Press Bureau Ltd.
P.O. Box 105, 107 Charterhouse Street
London EC1M 6AY, England

390. **Petroleum Engineer International.** 1929- . 13 issues/yr. $50.00/yr. ISSN 0164-8322.
>Harcourt Brace Jovanovich, Energy Publications Division
P.O. Box 1589
Dallas, TX 75221-1589

391. **Petroleum Frontiers.** 1983- . quarterly. $138.00/yr. ISSN 0740-1817.
>Petroleum Information Corporation
P.O. Box 2612
Denver, CO 80201

392. **Petroleum Intelligence Weekly.** 1962- . weekly. $850.00/yr. ISSN 0480-2160.
>Petroleum Intelligence Weekly
One Times Square Plaza
New York, NY 10036

393. **Petroleum Marketer.** 1933- . bimonthly. $18.00/yr. ISSN 0362-7799.
>McKeand Publications, Inc.
P.O. Box 507
West Haven, CT 06516

394. **Petroleum Outlook.** 1948- . monthly. $76.00/yr. ISSN 0031-6490.
>John S. Herold, Inc.
35 Mason Street
Greenwich, CT 06830

395. **Petroleum Review.** 1947- . monthly. £42.35/yr. (overseas); £30.25 (inland). ISSN 0020-3076.
>Institute of Petroleum
61 New Cavendish Street
London W1M 8AR, England

142 / 9—PERIODICALS

396. **Petroleum Times.** 1899- . monthly. $125.00/yr. ISSN 0263-3590.
Business Press International Ltd.
Quadrant House, The Quadrant
Sutton, Surrey SM2 5AS, England

397. **Pipe Line Industry.** 1954- . monthly. contact publisher for price. ISSN 0032-0145.
Gulf Publishing Co.
P.O. Box 2608
Houston, TX 77001

398. **Pipeline.** 1928- . 7 issues/yr. $18.00/yr. ISSN 0148-4443.
Oildom Publishing Company of Texas, Inc.
3314 Mercer Street
Houston, TX 77027

399. **Pipeline & Gas Journal.** 1859- . monthly. $40.00/yr. ISSN 0032-0188.
Harcourt Brace Jovanovich, Energy Publications Division
P.O. Box 1589
Dallas, TX 75221-1589

400. **Platt's Oilgram News.** 1923- . daily, weekdays. $927.00/yr. ISSN 0163-1284.
McGraw-Hill, Inc.
1221 Avenue of the Americas
New York, NY 10020

401. **Quaternary Research: An Interdisciplinary Journal.** 1970- . bimonthly. $134.00/yr. (U.S.); $155.50 (foreign). ISSN 0033-5894.
Academic Press
111 Fifth Avenue
New York, NY 10003

402. **Review of Scientific Instruments.** 1930- . monthly. $300.00/yr. ISSN 0034-6748.
American Institute of Physics
335 East 45th Street
New York, NY 10017

403. **Royal Astronomical Society Geophysical Journal.** 1958- . monthly. $450.00/yr. ISSN 0016-8009.
Blackwell Scientific Publications Ltd.
Osney Mead
Oxford OX2 0EL, England

404. **Science.** 1880- . weekly. $60.00/yr. (individuals); $98.00/yr. (institutions). ISSN 0036-8075.
American Association for the Advancement of Science
1333 H Street, NW
Washington, DC 20005

405. **Scientific American.** 1845- . monthly. $24.00/yr. ISSN 0036-8733.
Scientific American, Inc.
415 Madison Avenue
New York, NY 10017

406. **Sea Technology.** 1960- . monthly. $20.00/yr. ISSN 0093-3651.
Compass Publications, Inc.
1117 N. 19th Street, Suite 1000
Arlington, VA 22209

407. **Sedimentary Geology.** 1967- . 20 issues/yr. $395.00/yr. ISSN 0037-0738.
Elsevier Scientific Publishing Co., Journals Department
P.O. Box 211
1000 AE Amsterdam, The Netherlands

408. **Sedimentology.** 1952- . bimonthly. $184.00/yr. ISSN 0037-0746.
Blackwell Scientific Publications
Osney Mead
Oxford OX2 0EL, England

409. **Society of Petroleum Engineers Journal.** 1961-1985. bimonthly. ISSN 0197-7520.
Society of Petroleum Engineers
222 Palisades Creek Drive
Richardson, TX 75080

See: SPE Drilling Engineering
SPE Formation Evaluation
SPE Petroleum Engineering
SPE Reservoir Engineering

410. **Southwest Oil World.** 1954- . monthly, twice in October of even-numbered years. $24.00/yr. ISSN 0884-6219.
Hart Publications Inc.
1900 Grant Street, Suite 400, P.O. Box 1917
Denver, CO 80201-1917

411. **SPE Drilling Engineering.** 1986- . bimonthly. $17.50/yr. (members); $35.00/yr. (nonmembers). ISSN 0885-9744.
Society of Petroleum Engineers
222 Palisades Creek Drive
Richardson, TX 75080

412. **SPE Formation Evaluation.** 1986- . bimonthly. $17.50/yr. (members); $35.00/yr. (nonmembers). ISSN 0885-923X.
Society of Petroleum Engineers
222 Palisades Creek Drive
Richardson, TX 75080

144 / 9 – PERIODICALS

413. **SPE Production Engineering.** 1986- . bimonthly. $17.50/yr. (members); $35.00/yr. (nonmembers). ISSN 0885-9221.
    Society of Petroleum Engineers
    222 Palisades Creek Drive
    Richardson, TX 75080

414. **SPE Reservoir Engineering.** 1986- . bimonthly. $17.50/yr. (members); $35.00/yr. (nonmembers). ISSN 0885-9248.
    Society of Petroleum Engineers
    222 Palisades Creek Drive
    Richardson, TX 75080

415. **Tectonics.** 1982- . bimonthly, with an additional issue in January. $88.00/yr. (nonmembers). ISSN 0278-7407.
    American Geophysical Union
    2000 Florida Avenue, NW
    Washington, DC 20009

416. **Tectonophysics.** 1964- . 44 issues/yr. FL 2519/yr. ISSN 0040-1951.
    Elsevier Scientific Publishing Co.
    P.O. Box 330
    1000 AH Amsterdam, The Netherlands

417. **Well Servicing.** 1961- . bimonthly. contact publisher for price. ISSN 0043-2393.
    Workover/Well Servicing Publications, Inc.
    6060 North Central Expressway, Suite 538
    Dallas, TX 75206

418. **Western Oil World.** 1944- . monthly. $24.00/yr. ISSN 0043-3985.
    Hart Publications Inc.
    1900 Grant Street, Suite 400, P.O. Box 1917
    Denver, CO 80201-1917

419. **Work Boat.** 1943- . monthly. $20.00/yr. ISSN 0043-8014.
    H. L. Peace Publications
    1421 South Sheridan Road
    Tulsa, OK 74112

420. **World Oil.** 1916- . monthly, twice in February and August. contact publisher for price. ISSN 0043-8790.
    Gulf Publishing Co.
    P.O. Box 2608
    Houston, TX 77001

10 Professional and Trade Associations

This chapter contains a comprehensive list of petroleum-related associations. Associations or professional societies make a wealth of resources available to their members. In addition to the opportunities for contact between professionals in a similar field, many associations organize professional meetings; publish proceedings, newsletters, books, or periodicals; provide library services; maintain directories of the membership; and assist in career planning.

The American Petroleum Institute, Gas Processors Association, Institute of Petroleum (London), American Society for Testing and Materials, and National Association of Corrosion Engineers are associations that also are involved in establishing petroleum industry standards and recommended practices. Publications lists from these and other associations usually include standards and are good indicators of the scope of interests and activities of the associations.

Information for every entry in this chapter has been verified with the association. An asterisk (*) has been used to indicate that a membership directory for that association is included in chapter 6.

Alabama Natural Gas Association
300 Vestavia Office Park, Suite 2300
Birmingham, AL 35216
(205) 979-9492

Alabama Oilmen's Association
401 Interstate Park Drive
Montgomery, AL 36109
(205) 277-0786

146 / 10 – PROFESSIONAL AND TRADE ASSOCIATIONS

Alaska Oil & Gas Association
505 W. Northern Lights Blvd., Suite 219
Anchorage, AK 99503
(907) 272-1481

American Association for the Advancement of Science
1333 H Street, NW
Washington, DC 20005
(202) 326-6400

American Association of Petroleum Geologists*
P.O. Box 979
Tulsa, OK 74101
(918) 584-2555

American Association of Petroleum Landmen
777 Main St., Suite 1470
Fort Worth, TX 76102
(817) 335-2275

American Chemical Society
1155 16th St., NW
Washington, DC 20036
(202) 872-4600

American Gas Association
1515 Wilson Blvd.
Arlington, VA 22209
(703) 841-8400

American Geological Institute
4220 King St.
Alexandria, VA 22302
(703) 379-2480

American Geophysical Union
2000 Florida Ave.
Washington, DC 20009
(202) 462-6903

American Independent Refiners Association
114 3rd St., SE
Washington, DC 20003
(202) 543-8811

American Institute of Chemical Engineers
345 East 47th St.
New York, NY 10017
(212) 705-7338

American Institute of Professional Geologists
7828 Vance Dr., Suite 103
Arvada, CO 80003
(303) 431-0831

American Petroleum Credit Association
City Place, Suite 600
730 Hennepin Ave., S
Minneapolis, MN 55403
(612) 341-9646

American Petroleum Institute*
1220 L St., NW
Washington, DC 20005
(202) 682-8000

American Society for Nondestructive Testing
4153 Arlingate Plaza, Caller #28518
Columbus, OH 43228-0518
(800) 222-ASNT; in Ohio (800) NDT-OHIO

American Society for Photogrammetry
210 Little Falls St.
Falls Church, VA 22046-4398
(703) 534-6617

American Society for Testing and Materials
1916 Race St.
Philadelphia, PA 19103
(215) 299-5400

American Society of Lubrication Engineers
838 Busse Hwy.
Park Ridge, IL 60068
(312) 825-5536

American Society of Mechanical Engineers
345 East 47th St.
New York, NY 10017
(212) 705-7722

American Society of Petroleum Operations Engineers
P.O. Box 956
Richmond, VA 23207
(804) 271-0794 or (703) 768-4159

10 – PROFESSIONAL AND TRADE ASSOCIATIONS / 147

Arizona Petroleum Marketers Association
P.O. Box 26905
Phoenix, AZ 85068
(602) 942-9173

Arkansas Oil Marketers Association
P.O. Box 229
Little Rock, AR 72203
(501) 374-6293

Association for Women Geoscientists
2352 West Third Street
Los Angeles, CA 90057
(213) 383-1228

Association of American State Geologists
c/o R. Thomas Segall
Michigan Department of Natural
 Resources, Geological Survey Division
P.O. Box 30028
Lansing, MI 48917

Association of British Independent Oil Exploration Companies
c/o Premier Consolidated Oilfields Plc.
23 Lower Belgrave Street
London SW1W ONR, England
(01) 730-0752

Association of Crude Oil and Gas Producers
Wirtschaftsverband Erdöl-und-
 Erdgasgewinnung
Brühlstrasse 9
D-3000 Hannover, West Germany
0511-326016

Association of Desk & Derrick Clubs of North America
315 Silvey Bldg.
Tulsa, OK 74119
(918) 587-3076

Association of Earth Science Editors
c/o American Geological Institute
4220 King St.
Alexandria, VA 22302
(703) 379-2480

Association of Energy Professionals
521 5th Ave.
New York, NY 10175
(212) 757-6454

Association of Engineering Geologists
5313 Williamsburg Rd., P.O. Box 1068
Brentwood, TN 37027
(615) 377-3578

Association of Oil Pipe Lines
1725 K St., NW, Suite 1205
Washington, DC 20006
(202) 331-8228

Association of Oilwell Servicing Contractors
6060 North Central Expressway, Suite 538
Dallas, TX 75206
(214) 692-0771

Association of Petroleum Writers
c/o Alan Petzet
P.O. Box 1260
Tulsa, OK 74101
(918) 835-3161

Association of the German Petroleum Industry
Mineraloelwirtschaftsverband e.V.
Steindamm 71
2000 Hamburg 1, West Germany
040-28541

Association of United Kingdom Oil Independents
Windsor & Regent House, 83/89 Kingsway
London WC2B 6SD, England
(01) 404-5161

Association technique de l'industrie du gaz en France
62, rue de Courcelles
75008 Paris, France
(1)47 54 34 34

British Standards Institution
2 Park Street
London, W1A 2BS, England
(01) 629-9000

10 – PROFESSIONAL AND TRADE ASSOCIATIONS

California Gas Producers Association
626 Wilshire Blvd.
Los Angeles, CA 90017
(213) 626-4821

California Independent Oil Marketers Association
555 Capitol Mall, Suite 655
Sacramento, CA 95814
(916) 441-5166

California Independent Producers Association
12062 Valley View Street, Suite 201
Garden Grove, CA 92645
(714) 895-2525 or (213) 594-8639

Canada Gas Association
55 Scarsdale Road
Toronto, ONT M3B 2R3, Canada
(416) 447-6455

Canadian Petroleum Association
1500, 633 Sixth Avenue SW
Calgary, ALB T2P 2Y5, Canada
(403) 269-6721

Canadian Society of Exploration Geophysicists
229, 640-5 Ave. SW
Calgary, ALB T2P OM6, Canada

Clay Minerals Society
c/o Susan Wintsch
P.O. Box 2295
Bloomington, IN 47402

Coastal Engineering Research Council
American Society of Civil Engineers
c/o Cubit Engineering Ltd.
207 E. Bay Street, Suite 311
Charleston, SC 29401

Colorado-Wyoming-New Mexico Petroleum Marketers Association
4465 Kipling Street
Wheat Ridge, CO 80033
(303) 422-7805

Committee for Women in Geophysics
c/o Society of Exploration Geophysicists
P.O. Box 702740
Tulsa, OK 74170-2740
(918) 493-3516

Council of Petroleum Accountants Societies
P.O. Box 13117
Arlington, TX 76013
(817) 265-7971

Cushman Foundation for Foraminiferal Research
E-501 U.S. National Museum
Washington, DC 20560
(202) 357-2405

Domestic Petroleum Council Inc.
1627 K Street, NW, Suite 910
Washington, DC 20006
(202) 223-9321

Eastern Mineral Law Foundation
West Virginia University, College of Law
P.O. Box 6130
Morgantown, WV 26506-6130
(304) 293-2470

European Association of Exploration Geophysicists
Wassenaarseweg 22
2596 CH The Hague, The Netherlands
(+31 70) 453688

European Liquefied Petroleum Gas Association
Association Européenne des Gaz de Pétrole Liquifies
18 rue Beffroy
92200 Neuilly/Seine, France
(1) 747-2542

Florida Natural Gas Association
230 Pinewood Drive
Eustis, FL 32726
(904) 589-4753

Florida Petroleum Council
325 John Knox Road, F-210
Tallahassee, FL 32303
(904) 386-3641

10 – PROFESSIONAL AND TRADE ASSOCIATIONS / 149

Florida Petroleum Marketers Association
209 Office Plaza
Tallahassee, FL 32301
(904) 877-5178

French Association of Independent Petroleum Companies
Association Française des Independants de Pétrole
4 avenue Hoche
75008 Paris, France
(1) 42 27 50 80

French Association of Petroleum Engineers
Association Française des Techniciens du Pétrole
14 Avenue de la Grande-Armee
75017 Paris, France
(1) 380-5426

Gas Processors Association
1812 First National Bank Bldg.
Tulsa, OK 74103
(918) 582-5112

Gas Processors Suppliers Association
1812 First National Bank Bldg.
Tulsa, OK 74103
(918) 582-5112

Gas Research Institute
8600 W. Bryn Mawr Ave.
Chicago, IL 60631
(312) 399-8100

Geochemical Society
c/o Dr. B. Gregor
Department of Geological Sciences, Wright State University
Dayton, OH 45435
(513) 873-3455

Geological Society of America*
P.O. Box 9140, 3300 Penrose Pl.
Boulder, CO 80301
(303) 447-2020

Georgia Oilmen's Association
3300 NE Expressway, Suite 8-P
Atlanta, GA 30341
(404) 451-5916

Geoscience Information Society
c/o American Geological Institute
4220 King St.
Alexandria, VA 22302
(703) 379-2480

Idaho Oil Marketers Association
P.O. Box 172
Boise, ID 83701
(208) 344-1152

Illinois Oil & Gas Association
P.O. Box 788
Mount Vernon, IL 62864
(618) 242-2857

Illinois Petroleum Marketers Association
112 W. Cook Street, Box 1508
Springfield, IL 62705
(217) 544-4609

Independent Gasoline Marketers Council
1133 15th St., NW, Suite 1100
Washington, DC 20005
(202) 857-0220

Independent Lubricant Manufacturers Association
1055 Thomas Jefferson Street, NW, Suite 302
Washington, DC 20007
(202) 337-3470

Independent Oil & Gas Association of New York
Guaranty Building, 28 Church Street
Buffalo, NY 14202
(716) 856-4791

Independent Oil & Gas Association of West Virginia
22 Capitol Street
Charleston, WV 25301
(304) 344-9867

Independent Oilmen's Association of New England
25 Sea Breeze Lane
New Castle, NH 03854
(603) 436-8424

10 – PROFESSIONAL AND TRADE ASSOCIATIONS

Independent Petroleum Association of America
1101 Sixteenth Street, NW
Washington, DC 20036
(202) 857-4722

Independent Petroleum Association of Mountain States
1214 Denver Club Building, 518 17th Street
Denver, CO 80202
(303) 623-0987

Independent Terminal Operators Association
1015 15th Street, NW, Suite 1100
Washington, DC 20005
(202) 393-8535

Indiana Oil Marketers Association
8780 Purdue Road
Indianapolis, IN 46268
(317) 875-9057

Institute of Gas Technology
3424 S. State Street
Chicago, IL 60616
(312) 567-3650

Institute of Petroleum
61 New Cavendish Street
London W1M 8AR, England
(01) 636-1004

Interessengemeinschaft Mittelständischer Mineralölverbände
Achim-von-Arnim Strasse 24
D-5300 Bonn 1, West Germany
0228-23-11-81

International Association for Mathematical Geology
Kansas Geological Survey
1930 Constant Ave.
Lawrence, KS 66046
(913) 864-4991

International Association of Drilling Contractors*
P.O. Box 4287
Houston, TX 77210
(713) 578-7171

International Association of Geophysical Contractors
5335 West 48th Ave., Suite 400
Denver, CO 80212
(303) 458-8404

International Cooperative Petroleum Association
30 Montgomery St.
Jersey City, NJ 07302
(201) 451-8330

International Federation of Petroleum and Chemical Workers
P.O. Box 6603
Denver, CO 80206
(303) 388-9237

International Oceanographic Foundation
3979 Rickenbacker Causeway
Miami, FL 33149
(305) 361-5786

International Oil Scouts Association*
5000 E. Ben White Blvd., Suite 301
Austin, TX 78741
(512) 448-4088

International Palaeontological Association
c/o William A. Oliver, Jr.
U.S. Geological Survey
E-305 Natural History Building,
  Smithsonian Institution
Washington, DC 20560
(202) 343-3523

International Pipe Association
7811 N. Alpine Road
Rockford, IL 61111
(815) 654-1902

International Pipe Inspectors Association
4101 Oates Road
Houston, TX 77013
(713) 678-7071

International Union of Petroleum and Industrial Workers
8131 E. Rosecrans Ave.
Paramount, CA 90723
(213) 630-6232

10 – PROFESSIONAL AND TRADE ASSOCIATIONS / 151

Interstate Oil Compact Commission*
P.O. Box 53127
Oklahoma City, OK 73152
(405) 525-3556

Irish Offshore Services Association
Confederation House, Kildare Street
Dublin 2, Ireland
(01) 779801

Japan Petroleum Development Association
Keidanren Kaikan, 9-4 1-chome
Ohtemachi, Chiyoda-Ku
Tokyo 100, Japan
(03) 279-5841

Japan Petroleum Institute
17 F Shin-Aoyama East Building
1-1 1-chome, Minami-Aoyama
Minato-Ku, Tokyo 107, Japan
(03) 475-1235

Kansas Independent Oil & Gas Association
500 Broadway Plaza, 105 S. Broadway
Wichita, KS 67202
(316) 263-7297

Kansas Oil Marketers Association
804 Merchants National Bank Building
Topeka, KS 66612
(913) 233-9655

Kentucky Independent Gasoline Marketers Association
3906 Dupont Square S.
Louisville, KY 40207
(502) 897-7167

Kentucky Oil & Gas Association
218 W. 8th Street
Owensboro, KY 42301
(502) 683-1886

Kentucky Petroleum Marketers Association
612-A Shelby Street
Frankfort, KY 40601
(502) 875-3742

Liaison Committee of Cooperating Oil & Gas Associations
105 S. Broadway, Suite 500
Wichita, KS 67202
(316) 263-7297

Louisiana Association of Independent Producers and Royalty Owners
Box 4229
Baton Rouge, LA 70821
(504) 388-9525

Louisiana LP Gas Association
P.O. Box 10446
Jefferson, LA 70121
(504) 831-0321

Louisiana Oil Marketers Association
P.O. Box 1
Shreveport, LA 71161
(318) 221-4113

Maine Oil Dealers Association
No. 9, US Route 1, Box 536
Yarmouth, ME 04096
(207) 846-5113

Marine Technology Society
2000 Florida Avenue, NW, Suite 500
Washington, DC 20009
(202) 462-7557

Maryland Petroleum Council
60 West Street, Suite 403
Annapolis, MD 21401
(301) 269-1850

Michigan Oil & Gas Association
1610 Michigan National Tower, P.O. Box 15069
Lansing, MI 49801
(517) 487-1092

Michigan Petroleum Association
1200 Michigan National Tower
Lansing, MI 48933
(517) 487-9139

Mid-Atlantic Petroleum Distributors Association
198 Thomas Johnson Drive
Frederick, MD 21701
(301) 694-0808

10 – PROFESSIONAL AND TRADE ASSOCIATIONS

Mid-Continent Oil & Gas Association
711 Adams Office Building
Tulsa, OK 74103
(918) 582-5166

Mid-Continent Oil & Gas Association, Louisiana Division
333 Laurel Street, Suite 740
Baton Rouge, LA 70801
(504) 387-3205

Midwest Gas Association
1111 Douglas Drive N.
Minneapolis, MN 55422
(612) 544-8272

Midwest Petroleum Marketers Association
1585 Des Plaines Ave.
Des Plaines, IL 60018
(312) 298-5370

Mineralogical Society
2000 Florida Avenue NW
Washington, DC 20009-1277
(202) 462-6913

Mining & Metallurgical Society of America
275 Madison Ave., Suite 2301
New York, NY 10016
(212) 684-4150

Mississippi Petroleum Marketers Association
P.O. Box 257
Jackson, MS 39205-0257
(601) 353-1624

Missouri Oil Jobbers Association
238 E. High Street
Jefferson City, MO 65101
(314) 635-7117

Montana Western Oil Marketers Association
1217 Wilder Ave.
Helena, MT 59601
(406) 442-6647

Mountain States Legal Foundation
1200 Lincoln Street, Suite 600
Denver, CO 80203
(303) 861-0244

National Association of Corrosion Engineers*
1440 South Creek Drive
Houston, TX 77084
(713) 492-0535

National Association of Mineral Oil Enterprises
Uniti Bundesverband mittelstaendischer Mineralölunternehmen e.V.
Buchtstrasse 10
2000 Hamburg 76, West Germany
040-2270030

National Association of Royalty Owners
119 N. Broadway
Ada, OK 74820
(405) 463-0034

National Drilling Contractors Association
3008 Millwood Ave., P.O. Box 11187
Columbia, SC 29211
(803) 252-5646

National LP-Gas Association
1301 West 22nd St.
Oak Brook, IL 60521
(312) 986-4800

National Lubricating Grease Institute
4635 Wyandotte St.
Kansas City, MO 64112
(816) 931-9480

National Petroleum Council
1625 K St., NW
Washington, DC 20006
(202) 393-6100

National Petroleum Refiners Association
1899 L St., NW, Suite 1000
Washington, DC 20036
(202) 457-0480

National Stripper Well Association
P.O. Box 3373
Abilene, TX 79604
(915) 672-5225

Natural Gas Supply Association
1730 Rhode Island Ave., NW, Suite 200
Washington, DC 20036
(202) 331-8900

Nebraska Petroleum Marketers
1320 Lincoln Mall
Lincoln, NE 68508
(402) 474-6691

Netherlands Oil and Gas Exploration and Production Association
Nederlandse Olie En Gas Exploratie En Produktie Associatie
Bezuidenhoutseweg 29
2594 AC The Hague, The Netherlands
070-478871

New England Fuel Institute Inc.
20 Summer Street, P.O. Box 888
Watertown, MA 02172
(617) 924-1000

New Mexico Oil & Gas Association
P.O. Box 1864
Santa Fe, NM 87504-1864
(505) 982-2568

New Mexico Petroleum Marketers Association. *See Colorado-Wyoming-New Mexico Petroleum Marketers Association*

New York State Oil Producers Association
Hilcrest Ave.
Olean, NY 14760
(716) 372-2534

North Carolina Petroleum Marketers Association
7300 Glenwood Avenue, P.O. Box 30519
Raleigh, NC 27622
(919) 782-4411

North Texas Oil and Gas Association
1106 City National Bank Building
Wichita Falls, TX 76301
(817) 723-4131

Northwest Petroleum Association
2345 Rice Street, Suite 173
St. Paul, MN 55113
(612) 484-7227

Norwegian Petroleum Institute
(Norsk Petroleumsinstitutt)
Bygdoy Allé 8, 0262 Oslo 2, Norway
(02) 430050

Norwegian Petroleum Society
Norsk Petroleumsforening
P.O. Box 1897
Vika, 0124 Oslo 1, Norway
(02) 20-70-25

Offshore Oil Scouts Association
P.O. Box 6946
Metairie, LA 70009

Ohio Oil & Gas Association
P.O. Box 535
Granville, OH 43023
(614) 587-0444

Ohio Petroleum Marketers Association
50 West Broad Street, Suite 1130
Columbus, OH 43215
(614) 224-6271

Oil Investment Institute
P.O. Box 8293
Washington, DC 20024
(202) 543-3536

Oil Jobbers of Wisconsin
P.O. Box 8518
Madison, WI 53708
(608) 221-4771

Oil Marketers Association
417 S. Dearborn St., Room 210
Chicago, IL 60605
(312) 663-1250

Oklahoma Independent Petroleum Association
124 E. Fourth Street
Tulsa, OK 74103
(918) 584-1233

Oklahoma-Kansas Oil & Gas Association
Division, Mid-Continent Oil & Gas Association
700 Adams Building, 403 South Cheyenne
Tulsa, OK 74103
(918) 583-8505

154 / 10 – PROFESSIONAL AND TRADE ASSOCIATIONS

Oklahoma Oil Marketers Association
5115 N. Western
Oklahoma City, OK 73118
(405) 842-6625

Oregon Petroleum Marketers Association
3130 SW Ridgewood Road
Portland, OR 97225
(503) 292-1495

Organisation of Arab Petroleum Exporting Countries
P.O. Box 20501
Safat
Kuwait City, Kuwait

Pacific Coast Gas Association
1350 Bayshore Hwy., Suite 340
Burlingame, CA 94010
(415) 579-7000

Paleontological Research Institution
1259 Trumansburg Rd.
Ithaca, NY 14850
(607) 273-6623

Paleontological Society
c/o John J. Pojeta
U.S. Geological Survey, E-501 U.S. National Museum
Washington, DC 20560
(202) 343-5097

Panhandle Producers and Royalty Owners Association
Texas Commerce Bank Building, 2201 Civic Circle, Suite 404
Amarillo, TX 79109
(806) 352-5637

Pennsylvania Grade Crude Oil Association
c/o Pringle Powder Company
P.O. Box 201
Bradford, PA 16701
(814) 368-8171

Pennsylvania Natural Gas Associates
600 Gate House, Station Square
Pittsburgh, PA 15219
(412) 391-2397

Pennsylvania Oil & Gas Association
P.O. Box 180, 108 Main Street
Bradford, PA 16701
(814) 362-6722

Pennsylvania Petroleum Association
2 Shore Drive Office Center, Suite 121
Harrisburg, PA 17102
(717) 233-5838

Permian Basin Petroleum Association
P.O. Box 132
Midland, TX 79702
(915) 684-6345

Petroleum Association of Japan
No. 9-4, 1-Chome, Ohtemachi
Chiyoda-ku, Tokyo, Japan
(03) 279-3811

Petroleum Electric Power Association
c/o Benjamin B. Benigno
West Texas Utilities Co.
P.O. Box 841
Abilene, TX 79604
(915) 672-3251

Petroleum Equipment Institute
P.O. Box 2380
Tulsa, OK 74101
(918) 743-9941

Petroleum Equipment Suppliers Association
9225 Katy Frwy., Suite 401
Houston, TX 77024
(713) 932-0168

Petroleum Industry Association
Association d'Industriels du Pétrole
4 avenue Hoche
75008 Paris, France
(1) 763-0043

Petroleum Industry Research Foundation
122 E. 42nd St.
New York, NY 10168
(212) 867-0052

Petroleum Industry Security Council
P.O. Box 15392
Austin, TX 78761
(512) 454-3562

10 — PROFESSIONAL AND TRADE ASSOCIATIONS / 155

Petroleum Marketers Association of America
1120 Vermont Ave., NW, Suite 1130
Washington, DC 20005
(202) 331-1198

Petroleum Marketers of Iowa
P.O. Box 6156
Des Moines, IA 50309
(515) 244-6273

Petroleum Marketing Education Foundation
101 North Alfred Street, Suite 200
Alexandria, VA 22314
(703) 684-0000

Pipeline Contractors Association
4100 First City Center, 1700 Pacific Avenue
Dallas, TX 75201-4618
(214) 969-2700

Portable Drilling Rig Manufacturers Association
221 N. LaSalle St., Suite 2026
Chicago, IL 60601
(312) 346-1600

Rocky Mountain Mineral Law Foundation
Fleming Law Building, Campus Box 405,
  University of Colorado
Boulder, CO 80309
(303) 492-6545

Rocky Mountain Oil & Gas Association
345 Petroleum Building
Denver, CO 80202
(303) 534-8261

Seismological Society of America
2620 Telegraph Ave.
Berkeley, CA 94704
(415) 848-0954

Shippers Oil Field Traffic Association
Tulsa Chamber of Commerce
616 S. Boston
Tulsa, OK 74119
(918) 585-1201

Society of Economic Geologists
P.O. Box 571
Golden, CO 80402
(303) 236-5538

Society of Economic Paleontologists and Mineralogists
P.O. Box 4756
Tulsa, OK 74159-0756
(918) 743-9765

Society of Exploration Geophysicists
P.O. Box 702740
Tulsa, OK 74170-2740
(918) 493-3516

Society of Independent Gasoline Marketers of America
1730 K Street, NW, Suite 907
Washington, DC 20006
(202) 429-9333

Society of Independent Professional Earth Scientists
4925 Greenville Ave., Suite 170
Dallas, TX 75206
(214) 363-1780

Society of Petroleum Engineers*
SPE World Headquarters
P.O. Box 833836
Richardson, TX 75083-3836
(214) 669-3377

Society of Piping Engineers & Designers
One Main St.
Houston, TX 77002
(713) 221-8090

Society of Professional Well Log Analysts
6001 Gulf Frwy., Suite C-129
Houston, TX 77023
(713) 928-8925

Society of Rheology
American Institute of Physics
335 E. 45th St.
New York, NY 10017
(212) 661-9404

10 – PROFESSIONAL AND TRADE ASSOCIATIONS

South Carolina Oil Jobbers Association
1809 Gadsden Street, P.O. Box 64
Columbia, SC 29202
(803) 765-9570

South Dakota Petroleum Marketers
221 S. Central
Pierre, SD 57501
(605) 224-8606

Southeastern Independent Oil Marketers Association
6045 Barfield Road, NE, Suite 104
Atlanta, GA 30328
(404) 256-5541

Southern Gas Association
4230 LBJ Frwy., Suite 414
Dallas, TX 75244-5894
(214) 387-8505

Southwest Kansas Royalty Owners Association
209 East 6th Street, P.O. Box 250
Hugoton, KS 67951
(316) 544-4333

Southwestern Legal Foundation
P.O. Box 830707
Richardson, TX 75083-0707
(214) 690-2370

Spill Control Association of America
17117 West Nine Mile Rd., Suite 1040
Southfield, MI 48075
(313) 552-0500

Technical Association for Liquefied Petroleum Gas
Vereniging Technische Commissie, Vloeibaar Gas
c/o Th.M.J.F. van Thiel
Waalseweg 1
5711 BM Someren, The Netherlands
04937-4707

Tennessee Oil & Gas Association
5600 Brookwood Terrace
Nashville, TN 37205
(615) 356-4240

Tennessee Oil Marketers Association
430 Enos Reed Drive, P.O. Box 23256
Nashville, TN 37202
(615) 242-4377

Texas Independent Producers and Royalty Owners Association
1910 Interfirst Tower
Austin, TX 78701
(512) 477-4452

Texas LP-Gas Association
Box 9925
Austin, TX 78766
(512) 836-8620

Texas Oil Marketers Association
701 W. 15th Street
Austin, TX 78701
(512) 476-9547

Tulsa Association of Petroleum Landmen
P.O. Box 544
Tulsa, OK 74101

U.K. Offshore Operators Association Ltd.
3 Hans Crescent
London SW1X 0LN, England
(01) 589-5255

U.K. Petroleum Industry Association Ltd.
9 Kingsway
London WC2B 6XH, England
(01) 240-0289

VEG-Gasinstitut Nv
P.O. Box 137
Wilmersdorf 50
7300 AC Apeldoorn, The Netherlands
055-422922

Virginia Oil & Gas Association
P.O. Box 1837
Wise, VA 24293
(703) 328-3739

Virginia Petroleum Jobbers Association
Box 6344
Richmond, VA 23230
(804) 282-7534

10 – PROFESSIONAL AND TRADE ASSOCIATIONS / 157

Washington Oil Marketers Association
400 Dexter Avenue N.
Seattle, WA 98109
(206) 623-8733

West Central Texas Oil & Gas Association
P.O. Box 2332
Abilene, TX 79604
(915) 677-2469

West Virginia Oil & Natural Gas Association
P.O. Box 3231
Charleston, WV 25332
(304) 343-1609

West Virginia Petroleum Council
714 Atlas Building
Charleston, WV 25301
(304) 344-3609

West Virginia Petroleum Marketers Association
Suite 806 Atlas Building
Charleston, WV 25301
(304) 343-5885

Western Oil and Gas Association
727 W. 7th St., Suite 850
Los Angeles, CA 90017
(213) 627-4866

Western Petroleum Marketers Association
1800 SW Temple, Suite 300
Salt Lake City, UT 84115
(801) 484-8719

World Federation of Pipe Line Contractors Associations
4100 First City Center, 1700 Pacific Avenue
Dallas, TX 75201-4618
(214) 969-2700

11 Publishers

The following list of publishers has been provided for convenience in ordering materials included in this guide. Every attempt has been made to supply current addresses and telephone numbers. *Books in Print*,[1] *Publishers, Distributors, & Wholesalers of the United States*,[2] and *Publishers' International Directory*,[3] contain comprehensive lists of foreign and domestic publishers and may be helpful in updating the information in this chapter.

AAPG. *See American Association of Petroleum Geologists*

ABC-Clio
Riviera Campus, 2040 Alameda Padre Serra
P.O. Box 4397
Santa Barbara, CA 93103
(805) 963-4221

AGA. *See American Gas Association*

AGI. *See American Geological Institute*

American Association of Petroleum Geologists
P.O. Box 979
Tulsa, OK 74101-0979
(918) 584-2555

The American Gas Association
1515 Wilson Blvd.
Arlington, VA 22209
(703) 841-8400

American Geological Institute
4220 King St.
Alexandria, VA 22302
(703) 379-2480

---

[1]*Books in Print* (New York: R. R. Bowker, 1947-). LC 4-12648. ISSN 0068-0214.
[2]*Publishers, Distributors, & Wholesalers of the United States* (New York: R. R. Bowker, 1979-). ISSN 0000-0671.
[3]*Publishers' International Directory* (Munich, West Germany: K. G. Saur, 1962-). ISSN 0074-9877.

11 – PUBLISHERS / 159

American Petroleum Institute
1220 L St., NW
Washington, DC 20005
(202) 682-8375
or
156 William St., 4th Floor
New York, NY 10038
(212) 587-9660

American Society for Testing and Materials
1916 Race Street
Philadelphia, PA 19103
(215) 299-5400

Anchor Press/Doubleday
501 Franklin Avenue
Garden City, NY 11530
(516) 294-4561

API. *See American Petroleum Institute*

Arab Petroleum Research Center
7 Avenue Ingres
75016 Paris, France
524-3310

Alan Armstrong
1606 Jackson St.
Amarillo, TX 79102
(806) 374-1818

ASR Marketing
31 Bridge Street
Hertfordshire HP4 2EB, England
(04427)4677

ASTM. *See American Society for Testing and Materials*

Ballinger Publishing Co.
54 Church St., P.O. Box 281, Harvard Square
Cambridge, MA 02138
(617) 492-0670

Matthew Bender
P.O. Box 658
Albany, NY 12201
(800) 833-9844 or (800) 422-2022 (in NY)

Benn Business Information Services Ltd.
Union House, Eridge Rd.
Tunbridge Wells, Kent TN4 8HF, England

Benn Technical Books
209 High Street
Croydon, CR0 1QR, England

R. R. Bowker
1180 Avenue of the Americas
New York, NY 10036
(212) 764-5100

Burmass Publishing Co.
P.O. Box 1768
Midland, TX 79702
(915) 682-1782

Busby Associates, Inc.
576 South 23rd Street
Arlington, VA 22202
(703) 892-2888

Butterworth Publishers
10 Tower Office Park
Woburn, MA 01801
(617) 933-8260

Butterworth Scientific Ltd.
P.O. Box 63
Westbury House, Bury Street
Guildford, Surrey, GU2 5BH, England

R. W. Byram & Company
P.O. Drawer 1867
Austin, TX 78767
(512) 478-2551

Chemical Abstracts Service
2540 Olentangy River Road, P.O. Box 3012
Columbus, OH 43210
(614) 421-3600

H. Clarkson & Company Ltd.
12 Camomile St.
London, EC3A 7BP, England
(01) 283-8955

Commerce Clearing House
4025 W. Peterson Ave.
Chicago, IL 60646

11 – PUBLISHERS

DeGolyer and MacNaughton
One Energy Square
Dallas, TX 75206
(214) 368-6391

Domestic Petroleum Publishers. *See Eastern Petroleum Directory, Inc.*

Eastern Petroleum Directory, Inc.
370 Morris Ave.
Trenton, NJ 08611
(609) 393-4694

Éditions Olivier Lesourd
66, rue de la Rouchefoucauld
75009 Paris, France
281-4001

Éditions Technip
27 rue Ginoux
75737 Paris CEDEX 15, France

Elsevier Applied Science Publishers Ltd.
22 Rippleside Commercial Estate, Ripple Road
Barking, Essex IG11 0SA, England
(01) 595-2121

Elsevier Science Publishing Co.
52 Vanderbilt Ave.
New York, NY 10017
(212) 867-9040

Elsevier Scientific Publishing Co.
Postbus 211
NL-1000AH Amsterdam, The Netherlands
(020) 5803911

Enercom
65-67 avenue des Champs-Elysees
75008 Paris, France
225-7125

Energy Communications, Inc.
P.O. Box 1589
Dallas, TX 75221

Energy Economics Research Ltd.
7/9 Queen Victoria St.
Reading, Berkshire RG1 1SY, England
(0734)587680/89

Energy Industries Council
178-202 Great Portland Street
London W1N 6DU, England

Energy Publications
800 Davis Building, P.O. Box 1589
Dallas, TX 75221
(214) 748-4403

Engineering Information, Inc.
345 East 47th Street
New York, NY 10017-2387
(212) 705-7600

FT Business Information Ltd.
102 Clerkenwell Road
London EC1M 5SA, England
(01) 251 9321

Gale Research Co.
Book Tower
Detroit, MI 48226
(313) 961-2242

The Geological Society of America
P.O. Box 9140, 3300 Penrose Place
Boulder, CO 80301
(303) 447-2020

Geophysical Directory, Inc.
P.O. Box 13508
Houston, TX 77219
(713) 529-8789

Golden Bell Press
2403 Champa St.
Denver, CO 80205
(303) 572-1777

Government Printing Office. *See Superintendent of Documents*

Graham & Trotman Ltd.
Sterling House, 66 Wilton Rd.
London, SW1V 1DE, England
(01) 821-1123

Greene Dot, Inc.
11844 Rancho Bernardo Road, Suite 31A
San Diego, CA 92128
(619) 485-7237

GSA. *See The Geological Society of America*

11 – PUBLISHERS / 161

Gulf Publishing Co.
P.O. Box 2608
Houston, TX 77001
(713) 520-4444

Hart Publications, Inc.
P.O. Box 1917
Denver, CO 80201
(303) 837-1917

Her Majesty's Stationery Office
49 High Holborn
London WC1 6HB, England

Howell Publishing Co.
P.O. Box 27561
Houston, TX 77227
(713) 961-4432

Hunter Publishing Co.
950 Lee St.
Des Plaines, IL 60016
(312) 296-0770

Hydrocarbon Processing Company. *See Gulf Publishing Co.*

IADC. *See International Association of Drilling Contractors*

IED. *See Institute for Energy Development*

IEEE. *See Inspec IEEE*

IHRDC. *See International Human Resources Development Corporation*

IMCO Services
A Halliburton Company
2400 West Loop South, P.O. Box 22605
Houston, TX 77027
(713) 561-1393

Independent Petroleum Association of America
1101 Sixteenth St., NW
Washington, DC 20036
(202) 857-4722

Inspec IEEE
445 Hoes Lane
Piscataway, NJ 08854

Institute for Energy Development
P.O. Box 19243
Oklahoma City, OK 73144
(405) 691-4449

Institute for Scientific Information
3501 Market St.
Philadelphia, PA 19104
(215) 386-0100

Institute of Gas Technology
3424 S. State Street
Chicago, IL 60616
(312) 567-3650

Institution of Geologists Ltd.
Burlington House
Piccadilly, London W1V 9HG, England
(01) 734-0751

International Association of Drilling Contractors
P.O. Box 4287
Houston, TX 77210
(713) 578-7171

International Energy Agency
2, rue Andre-Pascal
75775 Paris CEDEX 16, France

International Human Resources Development Corporation
137 Newbury St.
Boston, MA 02116
(617) 536-0202

International Oil Scouts Association
5000 East Ben White Blvd., Suite 301
Austin, TX 78741
(512) 448-4088

Interstate Oil Compact Commission
P.O. Box 53127
Oklahoma City, OK 73152
(405) 525-3556

IPC Science and Technology Press
205 East 42nd Street
New York, NY 10017

ISI. *See Institute for Scientific Information*

162 / 11 — PUBLISHERS

Japan International Consultants, Ltd.
2-5-19 Sekimae
Musashinoshi, Tokyo, Japan

Libraries Unlimited, Inc.
P.O. Box 263
Littleton, CO 80160-0263
(303) 770-1220

Lloyd's of London Press Ltd.
Sheepen Place
Colchester, Essex C03 3LP, England
(0206) 69222 Ext. 259

Lloyd's Register of Shipping
71 Fenchurch Street
London EC3M 4BS, England
(01) 709-9166

Longman Group Ltd.
Westgate House, The High
Harlow, Essex CM20 1NE, England
(0279) 442601

Marlin Publications International
485 Fifth Ave.
New York, NY 10017
(212) 986-7752

Mathematical Geologists of the United States
c/o John C. Davis
Geologic Research Section, Kansas Geological Survey
Lawrence, KS 66045

McGraw-Hill Book Co.
1221 Avenue of the Americas
New York, NY 10020
(212) 512-2000

J. S. Metes & Company (PTE) Ltd.
Management Consultants
Tanglin Post Office Box 158
Singapore 9124, Republic of Singapore
737-3617

Midwest Oil Register
P.O. Box 700597
Tulsa, OK 74170
(918) 742-9925

Minerals Management Service, MS/640
Office of Offshore Information Services,
 U.S. Department of the Interior
Room 2070, Main Interior Building
Washington, DC 20240
(202) 343-3421

NACE. *See National Association of Corrosion Engineers*

National Association of Corrosion Engineers
1440 South Creek Drive
Houston, TX 77084
(713) 492-0535

National Supply Company
Pocket Reference
1455 West Loop South
Houston, TX 77027
(713) 960-5111

National Technical Information Service
5285 Port Royal Road
Springfield, VA 22161
(703) 487-4650

Nichols Publishing Co.
P.O. Box 96
New York, NY 10024
(212) 580-8079

Northern Miner Press Ltd.
7 Labatt Ave.
Toronto, ONT, Canada M5A 3P2
(416) 368-3481

Norwegian Petroleum Directorate
Lagardsveien 80, P.O. Box 600
N-4001 Stavanger, Norway
(04) 53-31-60

NTIS. *See National Technical Information Service*

OECD. *See Organization for Economic Co-operation and Development*

Offshore Information Literature
127 Eaton Manor, Eaton Gardens
Hove, Sussex BN3 3QD, England
(0273) 733241

11 — PUBLISHERS / 163

OGM Publishing Co., Inc.
One Poydras Plaza, 639 Loyola Ave.,
 Suite 1100
New Orleans, LA 70113
(504) 522-1100

Oil & Gas Directory
P.O. Box 13508
Houston, TX 77219
(713) 529-8789

Oil Daily
1401 New York Ave., Suite 500
Washington, DC 20005
(202) 662-0700

Oildom Publishing Company of Texas
3314 Mercer Street
Houston, TX 77027
(713) 622-0676

Oilfield Publications Ltd.
P.O. Box 11
Ledbury, Herefordshire HR8 1BN, England
0531-4563

Oklahoma Petroleum Directory
P.O. Box 700684
Tulsa, OK 74170
(918) 299-0194

OPEC. *See Organization of the Petroleum Exporting Countries*

Organization for Economic Co-operation and Development
2, rue Andre-Pascal
75775 Paris CEDEX 16, France

Organization of the Petroleum Exporting Countries
Obere Donaustrasse 93
A-1020 Vienna, Austria

Pasha Publications, Inc.
1401 Wilson Blvd., Suite 910
Arlington, VA 22209
(800) 424-2908

PennWell Publishing Co.
P.O. Box 1260
Tulsa, OK 74101
(918) 663-4220

Pergamon Press
Maxwell House, Fairview Park
Elmsford, NY 10523
(914) 592-7700

PETEX. *See Petroleum Extension Service*

PetroGuide
International Business Centre
90 Regent Street
London W1R 5PA, England
(01) 734 8466

Petroleum Equipment Institute
P.O. Box 2380
Tulsa, OK 74101
(918) 743-9941

Petroleum Extension Service
The University of Texas at Austin
BRC-2, 10100 Burnet Rd.
Austin, TX 78758
(512) 835-3163

Petroleum Information Corp.
P.O. Box 2612
Denver, CO 80201-2612

Petroleum News Southeast Asia Ltd.
6th Floor, 146 Prince Edward Road
W Kowloon, Hong Kong

Petroleum Publishers, Inc.
P.O. Box 129
Brea, CA 92621
(213) 691-1419

The Petroleum Publishing Co. *See PennWell Publishing Co.*

Phillips Petroleum Co.
16th Floor, Phillips Building
Bartlesville, OK 74004
(918) 661-4269

PIC. *See Petroleum Information Corp.*

Publishing Service United Nations
United Nations
New York, NY 10017

11 – PUBLISHERS

Resource Publications, Inc.
3210 Marquart
Houston, TX 77027
(713) 961-4191

Royal Ministry of Petroleum & Energy
P.O. Box 8148 Dep.
Oslo 1, Norway

Scarecrow Press
52 Liberty St., P.O. Box 656
Metuchen, NJ 08840
(201) 548-8600

SEG. *See Society of Exploration Geophysicists*

Society of Exploration Geophysicists
P.O. Box 702740
Tulsa, OK 74170-2740
(918) 493-3516

Society of Petroleum Engineers
SPE World Headquarters, P.O. Box 833836
Richardson, TX 75083-3836
(214) 669-3377

SPE. *See Society of Petroleum Engineers*

Spearhead Publications Ltd.
Rowe House, 55/59 Fife Rd.
Kingston upon Thames, Surrey KT1 1TA, England
(01) 549-5831

F. N. Spon
29 West 35th Street
New York, NY 10001

Strathclyde Regional Council
Industrial Development Unit
Strathclyde House, 20 India Street
Charing Cross, Glasgow G2 4PF, Scotland
(041) 227-3866

Superintendent of Documents
U.S. Government Printing Office
Washington, DC 20402
(202) 783-3238

Thomas Telford Ltd.
Telford House
P.O. Box 101
26/34 Old Street
London EC1P 1JH, England

Thompson-Wright Associates
P.O. Box 892
Golden, CO 80402

United Nations. *See Publishing Service United Nations*

U.S. Department of Energy, Technical Information Center. Documents available from *Superintendent of Documents* and from *National Technical Information Service.*

University Microfilms International
300 N. Zeeb Road
Ann Arbor, MI 48106

University of Texas at Austin. *See Petroleum Extension Service*

University of Tulsa
Division of Information Services
600 South College Ave.
Tulsa, OK 74104
(918) 592-6000

Van Nostrand Reinhold
135 West 50th St.
New York, NY 10020
(212) 265-8700

Otto Vieth Verlag
Neumann-Reichardt-Str. 34
P.O. Box 70 16 06
D-2000 Hamburg 70, West Germany

Washington Area Council of Engineering Laboratories
8811 Colesville Road, Suite 225
Silver Spring, MD 20910
(301) 588-8668

The Whico Atlas Co.
P.O. Box 25143
Houston, TX 77005
(713) 523-6673

Whitney Communications Corp. *See Oil Daily*

Whole World Publishing, Inc.
400 Lake Cook Rd., Suite 207
Deerfield, IL 60015
(800) 323-4305

John Wiley & Sons. *See Wiley Interscience*

John Wiley & Sons Ltd.
Baffins Lane
Chichester, Sussex PO19 1UD, England

Wiley Interscience
John Wiley & Sons
605 Third Avenue
New York, NY 10158
(212) 850-6418

Author/Title Index

Reference is to entry number. The letter "n" is used to designate citations to items found in annotations.

A-Z of Offshore Oil & Gas, 2nd ed., 32
AAPG Bulletin, 302
AAPG/CDS, Exploratory Well File, 249
AAPG Explorer, 303
Abstracts of Health and Environment Literature, 10n
Abstracts of Petroleum Substitutes Literature, 10n
Abstracts of Transportation and Storage Literature, 10n
Active Well Data On-line, 262
Adams, Neal J., 84
Africa-Middle East Petroleum Directory, 194n
AGI Data Sheets for Geology in the Field, Laboratory, and Office, 97
Agriculturals, 12n
AICHE Journal, 304
Air and Gas Drilling Manual, 82
L'Air Liquide, Division Scientifique, 48
Alaska Petroleum & Industrial Directory, 156
Alberta Oil Sands Index, 235n

Alberta Oil Sands Index (AOSI), 235
American Association of Petroleum Geologists Annual Report ..., 132
American Gas Association Monthly, 305
American Journal of Science, 306
American Petroleum Institute Directory, 133
Anderson, Gene, 98
Anderson, Kenneth E., 68
Angel, R. R., 95
Annuaire des statistiques mondiales de l'energie, 234
Annuaire européen du pétrole, 53
Annual Energy Review, 215n, 217n
AOSI. *See* Alberta Oil Sands Index
APEA Journal, 307
API Abstracts/Catalysts and Catalysis, 9
API Abstracts/Literature, 10, 236n
API Abstracts/Oilfield Chemicals, 11
API Abstracts/Patents, 12, 237n
API Master Well File, 250
APILIT, 236

APIPAT, 12n, 237
Arab Business Yearbook, 60
Arab Oil & Gas Directory, 61
ARGREP (Petroleum Argus Daily Market Report), 251
ARGUS (Petroleum Argus), 263
Armstrong Oil Directories Eastern United States, 157
Armstrong Oil Directories Gulf Coast, 158
Armstrong Oil Directories Rocky Mountain and Central United States, 159
Armstrong Oil Directories Texas and Southeastern New Mexico, 160
Arnould, Michel, 45
Arthur Young's Oil and Gas Federal Income Taxation, 22nd ed., 102
Ash, Lee, 76
Asia-Pacific/Africa-Middle East Petroleum Directory, 194
Asia-Pacific Petroleum Directory, 194n
Atomindex, 240n
Austin, Ellis H., 85
Author and Title Indexes to the Microfiche Collection (SPE Technical Papers), 23

Baker, Ron, 71, 93
Balachandran, Sarojini, 218
Banque de données du sous-sol français, 264
Basic Petroleum Data Book, 214
Bates, Robert L., 34, 34n, 36
Beck, Robert J., 228
Berger, Bill D., 68
Berry, Richard W., 28
BETC Crude Oil Analysis Data Bank, 265
BETC RI 78/14, 265n
Bibliography and Index of Geology, 1, 243n
Bibliography and Index of Geology Exclusive of North America, 1n
Bibliography and Index of North American Geology, 1n
Bibliography of Economic Geology, 242n
Bier, Robert A., Jr., 5

BIG. See Bibliography and Index of Geology
Biomass Abstracts, 240n
Bishop, Elna E., 131
Bland, William F., 126
BMR Journal of Australian Geology & Geophysics, 308
Brace, Gerald, 40
Bulletin of Canadian Petroleum Geology, 309
Burmass' Tex-Ok-Kan Oil Directory, 161

CA. See Chemical Abstracts
Canadian Journal of Earth Sciences, 310
Canadian Oil & Gas Handbook, 77
Canadian Oil Industry Directory, 184
Canadian Petroleum, 311
Catalogue: British Suppliers to the Oil, Gas, Petrochemical and Process Industries, 197
Chaballe, L. Y., 41
Champness, Michael, 229
Chemical Abstracts, 13
Chemical & Engineering News, 312
Chemical Engineering, 313
Chemical Engineering Progress, 314
Chemical Market Associates Petrochemical Market Reports. *See* CMAI
Chemical Products, 12n
China Offshore Oil Directory, 195
China Oil, 315
Christie, Helene Borgen, 105
Chryssostomidis, Marjorie, 7
Clays and Clay Minerals, 316
CMAI (Chemical Market Associates Petrochemical Market Reports), 252
Coal Abstracts, 240n
Composite Catalog of Oil Field Equipment & Services, 198
Computers & Chemical Engineering, 317
Computers & Geosciences, 318
Computers and Geotechnics, 319
Conservation—Transportation—Engineering—Storage, 12n
Continental Shelf, 103

AUTHOR/TITLE INDEX / 169

Coring and Core Analysis Handbook, 98
Corrosion, 320
Corrosion Abstracts, 14
Cost Estimating Manual for Pipelines and Marine Structures, 114
Costs and Indexes for Domestic Oil and Gas Field Equipment and Production Operations, 215n
Country Acreage and Activity Statistics, 285n
Crabbe, David, 31
CREW (Seismic Crew Count), 266
Crude Oil Pipeline Atlas of the United States and Canada, 115
Current Bibliography of Offshore Technology and Offshore Literature Classification, 2nd ed., 2
Current Energy Patents, 240n
Curtis, Doris M., 66

D & D Standard Oil Abbreviator, 2nd ed., 33
Davenport, Byron, 88
Davidson, R. L., 126
Desbrandes, Robert, 47
Desirable Energy Future, 217n
Development of the Oil and Gas Resources of the United Kingdom, 57
DeWitt (DeWitt Petrochemical Newsletters), 253
DeWitt Petrochemical Newsletters. *See* DeWitt
Dictionary of Energy Technology, 24
Dictionary of Geological Terms, 3rd ed., 34
Dictionary of Geosciences, 39
Dictionary of Petroleum Technology, 40
Dictionary of Petroleum Terms, 3rd ed., 25
Dictionnaire technique du pétrole, 40
Dietrich, R. V., 97
Directory Interstate Oil Compact Commission and State Oil and Gas Agencies, 134
Directory of Certified Petroleum Geologists. *See 1985 Directory of Certified Petroleum Geologists*
Directory of Geophysics Education, 162
Directory of Geoscience Departments United States & Canada, 163
Directory of North American Geoscientists Engaged in Mathematics, Statistics and Computer Applications, 135
Directory of Oil Well Drilling Contractors, 143
Directory of Pipe Line Companies and Pipe Line Contractors, 144
Directory of Producers and Drilling Contractors: California, 164
Directory of Producers and Drilling Contractors: Louisiana, Mississippi, Arkansas, Florida, Georgia, 165
Directory of Producers and Drilling Contractors: Oklahoma, 166
Directory of Producers and Drilling Contractors: Rocky Mountain Region, Williston Basin, Four Corners, New Mexico, 167
Directory of Producers and Drilling Contractors: Texas, 168
DIS. *See* Drilling Information Services
Doherty, W. T., 75
DPDS. *See* Dwight's Petroleum Data System
DRI Natural Gas, 267
DRI Oil and Gas Drilling, 268
DRI Oil Company, 269
DRI U.S. Energy, 270
Drill Bit. *See Southwest Oil World*
Drilling Analysis Data On-Line, 271
Drilling Contractor, 321
Drilling Data Handbook, 83
Drilling Engineering, 84
Drilling Engineering Handbook, 85
Drilling Information Services (DIS), 254
Drilling Practices Manual, 2nd ed., 86
Drilling, The Wellsite Publication, 322
Dutro, J. T., 97
Dwight's On-line System, 272
Dwight's Petroleum Data System (DPDS), 255

Earth and Planetary Science Letters, 323
Eastern Petroleum Directory, 169

170 / AUTHOR/TITLE INDEX

EBIB. *See* Energy Bibliography & Index (EBIB)
Eckel, Edwin B., 131
ECOMINE, 238
Ecomine Minière, 238n
Ecomine, Revue de presse, 238n
Economic Geology and The Bulletin of the Society of Economic Geologists, 324
EDB. *See* Energy Data Base
EDPRICE (Lundberg Survey Energy Detente International Price and Tax Series), 273
EIA Publications Directory, 215
Electronic Markets and Information Systems Inc. *See* EMIS
Electronic Rig Stats (ERS), 256
Elsevier's Oil and Gas Field Dictionary in Six Languages, 41, 43n
EMIS (Electronic Markets and Information Systems Inc.), 274
Encyclopedia of Atmospheric Sciences and Astrogeology, 46n
Encyclopedia of Beaches and Coastal Environments, 46n
Encyclopedia of Earth Sciences Series, 46
Encyclopedia of Geochemistry and Environmental Sciences, 46n
Encyclopedia of Geomorphology, 46n
Encyclopedia of Mineralogy, 46n
Encyclopedia of Oceanography, 46n
Encyclopedia of Paleontology, 46n
Encyclopedia of Sedimentology, 46n
Encyclopedia of Soil Science, 46n
Encyclopedia of Well Logging, 47
Encyclopedia of World Regional Geology, 46n
Encyclopedic Dictionary of Exploration Geophysics, 2nd ed., 35
Encyclopedie des gaz, 48
Energy Abstracts for Policy Analysis, 240n
Energy and the Environment, 240n
Energy Bibliography and Index, 3, 239n
Energy Bibliography & Index (EBIB), 3n, 239
Energy Daily, 325
Energy Data Base (EDB), 240
Energy Decade 1970-1980, 216, 217n
Energy Detente, 273n

Energy Exploration & Exploitation, 326
Energy Fact Book, 217
Energy Handbook, 2nd ed., 64
Energy Historical Database, 275
Energy Index, 241n
Energy Information Abstracts, 241n
Energy Information Guide, 4
Energy Meetings & Trade Shows Worldwide Directory, 145
Energy Research Abstracts, 240n
Energy Statistics, 218
ENERGYLINE, 241
Engineering Index, 15n
Engineering Index Annual, 15
Engineering Index Monthly and Author Index, 15n
English-Spanish and Spanish-English Glossary of the Petroleum Industry, 2nd ed., 42
Enhanced Recovery Week, 327
Enhanced Recovery Week EOR Project Sourcebook, 121
Environment Abstracts/Index, 241n
EOS, Transactions, American Geophysical Union, 328
Erdoel-Erdgas Zeitschrift, 382
ERS. *See* Electronic Rig Stats
Essertier, Edward P., 219
European Offshore Oil & Gas Yearbook and Directory, 54
European Petroleum Directory, 187
European Petroleum Year Book, 53
European Sources of Scientific and Technical Information, 5th ed., 188

Fact Sheet: The Norwegian Continental Shelf, 55
Fairbridge, Rhodes W., 46
Federal Offshore Statistics, 219
Financial Times Energy World, 220
Financial Times Oil and Gas International Yearbook, 146
"Financial Times" Who's Who in World Oil and Gas, 136
Foose, R. M., 97
Foreign Scouting Service, 285n
Formulaire du foreur, 83n
Fossil Energy Update, 240n
Fox, E. C., 217

Frick, Thomas C., 123
Fuel, 329
Fuel Abstracts and Current Titles, 16n
Fuel and Energy Abstracts, 16
Fundamentals of Drilling, 87
Fundamentals of Petroleum, 2nd ed., 65
Fusion Energy Update, 240n

Gas Abstracts, 17
Gas Data Book, 221
Gas Directory and Who's Who, 137
Gas Encyclopaedia, 48
Gas Facts, 221n, 222
General Index to Publications of the Society of Petroleum Engineers of AIME, 18
GEOARCHIVE, 242
Geochemistry International, 330
Geochimica et cosmochimica acta, 331
Geokhimiya, 330
Geologic Reference Sources, 2nd ed., 5
Geological Society of America Bulletin, 332
Geological Society of America Membership Directory, 138
Geologist's Directory, 3rd ed., 189
Geologiya i Geofizika, 333
Geologiya Nefti i Gaza, 334
Geology, 335
Geophysical Directory, 170
Geophysical Prospecting, 336
Geophysics, 337
GeoRef, 1n, 243
Geoscience Canada, 338
Geoscience Documentation, 242n
Geoscience Software Directory for the IBM PC & Compatibles, 199
Geotimes, 339
Geotitles, 242n
Gerolde, Steven, 27n
Gilpin, Alan, 24
Giuliano, Francis A., 67
Glosario de la industria petrolera, 42
Glossary of Geology, 2nd ed., 34n, 36
Glossary of Petroleum Industry Equipment Terms & Phrases, 44n
Guide du pétrole gaz—petrochimie, 190
Guide Offshore: French Offshore Yearbook, 56

Guide to Petroleum Statistical Information, 223
Guide to the Energy Industries, 6
Gulf Coast Oil Directory, 171
Gulf Coast Oil World, 340

Handbook of Drilling Practices, 88
Handbook of Energy Technology and Economics, 78
Handbook of Oceanic Pipeline Computations, 116
Handbook of Oil Industry Terms and Phrases, 4th ed., 26
Handbook of Pipeline Engineering Computations, 117
Handbook of Valves, Piping and Pipelines, 118
Handbook on Petroleum Land Titles, 104
Hankinson, R. L., 51
Hankinson, R. L., Jr., 51
Heavy Oil/Enhanced Recovery Index, 244n
Heavy Oil/Enhanced Recovery Index (HERI), 244
HERI. See Heavy Oil/Enhanced Recovery Index
Historical Well Data On-line, 276
Hobson, G. D., 99
Houston Oil Directory, 172
How to (Try to) Find an Oil Field, 66
HUGHES (Hughes Rotary Drilling Rig Reports), 277
Hughes Rotary Drilling Rig Reports. See HUGHES
Hughes Rotary Rig Reports, 277n
Hydrocarbon Processing, 341
Hydrocarbon Processing Catalog, 200

IADC Drilling Manual, 89
IADC Ninth Edition Drilling Manual, 94n
ICIS (Independent Chemical Information Services), 257
Illustrated Petroleum Reference Dictionary, 3rd ed., 27
Imported Crude Oil and Petroleum Products Report, 278n

IMPORTS (Imports of Crude Oil and Petroleum Products), 278
Imports of Crude Oil and Petroleum Products. *See* IMPORTS
Independent Chemical Information Services. *See* ICIS
INFOIL 2, 260
International Association of Drilling Contractors Membership Directory, 139
International Directory of Energy Meetings & Trade Shows, 145n
International Energy Annual, 215n
International Journal of Multiphase Flow, 342
International Oil and Gas Development, 49
International Oil Newsletter, 261n
International Oil Scouts Association Directory, 147
International Petroleum Abstracts, 19
International Petroleum Encyclopedia, 50
Interstate Oil Compact Commission, 134
Introduction to Oil and Gas Technology, 2nd ed., 67
Introduction to Petroleum Geology, 2nd rev. ed., 99
Introduction to Petroleum Production, 122
Isotope Geoscience, 343

Jackson, Julia A., 34, 34n, 36
Japan Petroleum Industry Yearbook, 59
Japan Petroleum Weekly, 344
Jenkins, Gilbert, 79, 229
Joint Association Survey on Drilling Costs, 224
Journal canadien des sciences de la terre, 310
Journal of Canadian Petroleum Technology, 345
Journal of Colloid and Interface Science, 346
Journal of Energy Resources Technology, 347
Journal of Engineering for Industry, 348
Journal of Engineering Materials and Technology, 349

Journal of Geochemical Exploration, 350
Journal of Geology, 351
Journal of Geophysical Research: Oceans and Atmospheres, 352
Journal of Geophysical Research: Solid Earth and Planets, 353
Journal of Geophysics, 354
Journal of Geotechnical Engineering, 355
Journal of Paleontology, 356
Journal of Petroleum Geology, 357
Journal of Petroleum Technology, 18n, 142n, 358
Journal of Physical Chemistry, 359
Journal of Sedimentary Petrology, 360
Journal of Structural Engineering, 361
Journal of Structural Geology, 362
Journal of the Acoustical Society of America, 363
Journal of the American Oil Chemists' Society, 364
Journal of the International Association for Mathematical Geology, 365
Journal of Transportation Engineering, 366

Kennedy, John L., 87
Kent Canadian Retail Gasoline Volume. *See* KENTV
KENTV (Kent Canadian Retail Gasoline Volume), 279

Laing, William E., 162
Land Drilling & Oilwell Servicing Contractors Directory, 173
Landman's Encyclopedia, 2nd ed., 51
Langenkamp, Robert D., 26-27
Latin American Petroleum Directory, 193
Leffler, William L., 128
Lilly, Willene Jackson, 74
Liquefied Petroleum Gas. *See* LPGAS
Liscom, William L., 216
Lloyd's Maritime Directory, 201
Loftness, Robert L., 64
Log Analyst, 367
London Oil Reports. *See* LOR
LOR (London Oil Reports), 280
LPGAS (Liquefied Petroleum Gas), 281

Lundberg Survey Energy Detente International Price and Tax Series. *See* EDPRICE
Lundberg Survey Retail Prices. *See* RETAIL
Lundberg Survey Share of Market. *See* SOM
Lundberg Survey Share of Market Summary. *See* SOMSUM
Lundberg Survey Wholesale Prices and Moves. *See* WHOLESALE
Lyons, William C., 82

MacTiernan, Brian, 101
Malinowsky, H. Robert, 8
Marine and Petroleum Geology, 368
Marine Geology, 369
Marine Geotechnology, 370
Marine Technology, 371
Marks, Alex, 116-17
Masuy, L., 41
McBride, Richard, 31
McPhater, Donald, 101
Medel, Otto, 120
MER (Monthly Energy Review), 282
Meyers, Charles J., 37
Meyers, Robert A., 78
Midcontinent Petroleum Directory, 174
Midwest Oil Register Directory of Pipe Line Companies and Pipe Line Contractors, 144n
Midwest Oil Register Directory of Producers and Drilling Contractors Louisiana, Mississippi, Arkansas, Florida, Georgia, 165n
Midwest Oil Register Directory of Producers and Drilling Contractors, Texas, 168n
Midwest Oil Register Oil Directory of Alaska, 176n
Midwest Oil Register Oil Directory of Canada, 186n
Midwest Oil Register Oil Directory of Foreign Companies Outside the U.S.A. and Canada, 148n
Midwest Oil Register Oil Directory of Houston, Texas, 177n
Midwest Oil Register Oil Directory of Producers and Drilling Contractors California, 164n

Midwest Oil Register Oil Directory of Producers and Drilling Contractors of Oklahoma, 166n
Midwest Oil Register Oil Directory of Producers and Drilling Contractors of Rocky Mountain Region, Williston Basin, Four Corners and New Mexico, 167n
Midwest Oil Register Oil Well Drilling Contractors, 143n
Miller's Oil and Gas Federal Income Taxation, 102n
Modern Petroleum, 2nd ed., 68
Monthly Energy Review, 217n, 282n
Monthly Energy Review. *See* MER
Monthly Statistical Report, 299n
Moore, Preston L., 86
Mosburg, Lewis G., Jr., 104
Moureau, Magdeline, 40
Mud Equipment Manual, 90
Mudfacts Engineering Handbook, 91
Myers, Arnold, 2

NACE Membership Directory, 140
National Petroleum News, 52n, 372
National Petroleum News Fact Book, 52
Nature, 373
Nelson, Don, 206
Nelson, W. L., 127
New Scientist, 374
1985 Directory of Certified Petroleum Geologists, 141
Noroil, 375
North Sea and North 62 Degrees Atlas, 108
North Sea Environmental Guide, 109
North Sea Oil & Gas Directory, 191
North Sea Platform Guide, 110
Norwegian Petroleum Guide, 105

Ocean Industry, 376
Offshore, 377
Offshore Abstracts, 20
Offshore Contractors & Equipment Directory, 202
Offshore Drilling Register, 203
Offshore Engineer, 378
Offshore Frontiers, 111
Offshore Magazine, 202n

Offshore Oil & Gas Yearbook, 58
Offshore Petroleum Engineering, 7
Offshore Service Vessel Register, 204
Offshore Services + Equipment Directory, 205
OIL, 245
Oil and Energy Trends Statistics Review, 225
Oil & Gas: The Production Story, 71
Oil & Gas Directory, 175
Oil and Gas Field Code Master List, 69
Oil & Gas Journal, 183n, 194n, 226n, 379
Oil & Gas Journal Data Book, 226
Oil & Gas Journal Energy Database, 283
Oil and Gas Pocket Reference 1984, 70
Oil & Gas Producing Industry in Your State, 227
Oil and Gas Terms, 6th ed., 37
Oil and Oilfield Equipment and Service Companies Worldwide, 206
Oil & Petrochemical Pollution, 380
Oil Daily, 381
Oil Directory of Alaska, 176
Oil Directory of Canada, 186
Oil Directory of Foreign Companies Outside the U.S.A. and Canada, 148
Oil Directory of Houston, Texas, 177
Oil Economists' Handbook 1985, 79
Oil Gas European Magazine, 382
Oil-Gas-Marine Directory, 149
Oil Industry Outlook for the United States, 228
Oil Price Information Service Newsletter, 288n
Oil Property Evaluation, 2nd ed., 80
Oil Tanker Databook 1985, 229
Oilman, 383
OILS Database, 258
Oilweek, 384
Oklahoma Petroleum Directory, 178
Olszewski, M., 217
Onshore/Offshore Oil & Gas Multilingual Glossary, 43
OPEC Facts & Figures, 230
OREDA Offshore Reliability Data Handbook, 112
Organic and Petroleum Chemistry for Nonchemists, 125

Pace Synthetic Fuels Report, 385
Pacific Coast Oil Directory, 179
Pacific Oil World, 386
Page, John S., 114
Palaeogeography Palaeoclimatology Palaeoecology, 387
Palaeontology, 388
Palmer, Susan Reeves, 29
PASCAL-FOLIO, 246n
PASCAL-GEODE, 246
PASCAL-THEMA Sciences de la Terre, 246n
P/E News, 22n, 247
Permian Basin Oil Directory, 180
Permit Data On-line, 284
Petroconsultants Exploration Database, 285
Petroconsultants International Drilling, 286
Petroconsultants News Database, 261
Petroflash (Petroflash! Crude and Product Reports), 287
PetroGuide, 231
Petroleo moderno, 68n
Petroleum Abstracts, 21, 248n
Petroleum Argus, 251n
Petroleum Argus. *See* ARGUS
Petroleum Argus Daily Market Report. *See* ARGREP
Petroleum Dictionary, 28
Petroleum Drilling Equipment, Terms & Phrases English-Spanish, Spanish-English, 44
Petroleum Economist, 389
Petroleum/Energy Business News Index, 22, 223n
Petroleum Engineer International, 72n, 390
Petroleum Engineering and Technology Schools, 150
Petroleum Engineering Schools Book 1985-86 Academic Year, 150
Petroleum Engineer's Continuous Tables Fieldbook, 72
Petroleum Equipment Directory, 207
Petroleum Frontiers, 391
Petroleum Handbook, 6th ed., 73
Petroleum Independent, 227n
Petroleum Industry Glossary, 29
Petroleum Industry Yellow Pages, 208
Petroleum Intelligence Weekly, 289n, 392

AUTHOR/TITLE INDEX / 175

Petroleum Intelligence Weekly. *See* PIW
Petroleum Law Guide 1985, 106
Petroleum Marketer, 393
Petroleum Marketing Monthly, 215n, 292n
Petroleum Marketing Monthly. *See* PMPRICE
Petroleum Measurement Tables, 81
Petroleum Outlook, 394
Petroleum Processes, 12n
Petroleum Processing Handbook, 126
Petroleum Production Handbook, 123
Petroleum Refinery Engineering, 4th ed., 127
Petroleum Refining and Petrochemicals Literature Abstracts, 10n
Petroleum Refining for the Non-technical Person, 128
Petroleum Review, 395
Petroleum Secretary's Handbook, 2nd ed., 74
Petroleum Software Directory, 209
Petroleum Substitutes, 12n
Petroleum Supply Annual, 215n
Petroleum Times, 396
Petroleum Training Directory, 210
PetroScan, 288
Phillips 66 Glossary, 30
Pipe Line Industry, 397
Pipe Line Rules of Thumb Handbook, 119
Pipeline, 398
Pipeline & Gas Journal, 399
Pipeline Annual Directory and Equipment Guide, 151
PIW (Petroleum Intelligence Weekly), 289
Platt's Global Alert, 290
Platt's Oil Prices Databank, 291
Platt's Oilgram News, 400
PMPRICE (Petroleum Marketing Monthly), 292
Pocket Guide for Mud Technology, 92
Polymers, 12n
Practical Petroleum Engineers' Handbook, 5th ed., 75
Practical Piping Handbook, 120
Primary and Specialty Products, 12n
Primer of Oilwell Drilling, 4th ed., 93
Primer of Oilwell Service and Workover, 3rd ed., 124
Production Data On-line, 293

QOS (Quarterly Oil Statistics), 294
Quarterly Oil and Gas Statistics, 232
Quarterly Oil Statistics. *See* QOS
Quarternary Research, 401

Register of Offshore Units, Submersibles & Diving Systems, 211
Resume: The Complete Annual Review of Oil and Gas Activity in the United States, 62
RETAIL (Lundberg Survey Retail Prices), 295
Review of Scientific Instruments, 402
Richardson, Jeanne M., 8
RIGS Database, 259
Rocky Mountain Petroleum Directory, 181
Royal Astronomical Society Geophysical Journal, 403

Salem, S., 41
Schmerling, Louis, 125
Science, 404
Science and Engineering Literature, 3rd ed., 8
Scientific American, 405
Sea Technology, 406
Sedimentary Geology, 407
Sedimentology, 408
Seismic Crew Count. *See* CREW
Sheriff, Robert E., 35
SI Drilling Manual, 94
Skinner, D. R., 122
Society of Petroleum Engineers Annual Technology Review ..., 142
Society of Petroleum Engineers Journal, 18n, 409
Society of Petroleum Engineers Publications Style Guide, 130
Society of Petroleum Engineers Technical Papers, 23
Solar Energy Update, 240n
SOM (Lundberg Survey Share of Market), 296
SOMSUM (Lundberg Survey Share of Market Summary), 297
South East Asia Oil Directory, 196
Southwest Oil World, 410
Southwest Petrodata Report, 182

176 / AUTHOR/TITLE INDEX

SPE Drilling Engineering, 411
SPE Formation Evaluation, 412
SPE Production Engineering, 413
SPE Reservoir Engineering, 414
Standard Definitions for Petroleum Statistics, 38
Statistiques trimestrielles du pétrole et du gas naturel, 232
Strathclyde Oil Register, 192
Subject and Fiche Number Indexes (SPE Technical Papers), 23
Subject Collections, 6th ed., 76
Subsea, 111n
Suggestions to Authors of the Reports of the United States Geological Survey, 6th ed., 131

Tanker Register, 113
Technical Data Book—Petroleum Refining, 4th ed., 129
Tectonics, 415
Tectonophysics, 416
Termes pétroliers dictionnaire anglais-français, 45
Texas Oil and Gas Handbook, 107
Thomann, Arthur E., 44
Thompson, Robert S., 80
Tiratsoo, E. N., 99
TULSA, 21n, 248
Tver, David F., 28
Twentieth Century Petroleum Statistics, 233

Undersea Vehicles Directory, 212
Union List of Geologic Field Trip Guidebooks of North America, 3rd ed., 100
UK Offshore Oil and Gas Directory, 54n
U.S. Crude Oil, Natural Gas, and Natural Gas Liquids Reserves, 215n
United States Department of Energy. See USDOE
U.S. Energy Industry Yearbook, 63
U.S.A. Oil Industry Directory, 183
U.S.A. Oilfield Service, Supply and Manufacturers Directory, 213
Universal Conversion Factors, 27n

USDOE (United States Department of Energy), 298

Vandenberghe, J. P., 41
Volume Requirements for Air & Gas Drilling, 95

WACEL Driller's Guide, 96
Ward, Dederick C., 5
Waring, R. H., 118
Watznauer, Adolf, 39
Weber, David R., 4
Weekly North West Europe, 261n
Weekly Petroleum Status Report, 215n
Weekly Statistical Bulletin, 299n
Weekly Statistical Bulletin, 299
Well Completions Database, 300
Well Servicing, 417
Well-site Geologist's Handbook, 101
Western Oil World, 418
Wheeler, Marjorie W., 5
Whitehead, Harry, 32
Whole World Oil Directory, 152
WHOLESALE (Lundberg Survey Wholesale Prices and Moves), 301
Who's Drilling, 258n, 259n
Williams, Alan, 24
Williams, Howard R., 37
Winston, W. P., 106
Work Boat, 419
World Energy Book, 31
World Energy Industry, 217n
World Energy Supplies in Selected Years, 234n
World Oil, 420
Worldwide Offshore Rigfinder, 256n
Worldwide Petrochemical Directory, 153
Worldwide Pipeline & Contractors Directory, 154
Worldwide Refining and Gas Processing Directory, 155
Worobec, Alexandra, 77
Wörterbuch Geowissenschaften, 39n
Wright, John D., 80

Yearbook of World Energy Statistics, 234

Zaba, Joseph, 75
Zubini, Fabio, 45

Subject Index

Reference is to entry number. The letter "n" is used to designate citations to items found in annotations. The letter "p" is used to designate occasional references to pages.

Abbreviations, 24n, 25n, 27n, 29n, 32n, 33, 101n, 130n
Abstracting services. *See* Abstracts
Abstracts, 10, 17, 19, 21, 235n, 236n, 238n, 244n, 248n. *See also* Indexes
   chemicals, 9, 11, 13
   corrosion, 14
   energy, 16, 240n
   engineering, 15n
   geology, 246n
   offshore industry, 20, 260n
   patents, 12, 237n
Accounting, 37n, 80n, 102n, 146n
Africa, 208n
   companies, 61n, 153n, 194
   patents, 237n
   production reports, 194n
   rig information, 277n
Agricultural, 12n, 156n
Alabama
   companies, 158n, 171n
   products, 208n

Alaska
   companies, 156, 176, 179n
   databases, 266n
   products, 208n
Alberta, 235, 255n. *See also* Canada
Alternate energy sources. *See* Coal; Energy alternatives; Geothermal energy; Hydroelectric power; Nuclear energy; Oil shale industry; Solar energy; Synfuels; Wind power
Annual reports, 132, 146n, 269n
Arab countries. *See* Middle East
Arabic-language sources, 41n, 93n. *See also* Middle East
Arizona
   companies, 159n
Arkansas
   companies, 158n, 165, 174n
   products, 208n
Artificial recovery. *See* Development; Enhanced recovery

179

180 / SUBJECT INDEX

Asia. *See also* China; Japan
  companies, 153n, 194, 196
  production reports, 194n
  rig information, 277n
Assets. *See* Investments
Associations, pp. 145-57
  Canada, 33n, 184n
  Europe, 54n, 56n, 58n, 137n, 190n, 191n
  international, 33n, 50n, 139, 146n, 147, 201n
  United States, 33n, 52n, 63n, 70n, 132, 133n, 135n, 138n, 140n, 142n, 169n, 174n, 179n, 181n, 183n
Atlases, 31. *See also* Maps
  offshore, 108
  pipeline, 115
Australia
  drilling, 258n, 259n

Banking, 6n, 60n, 63n. *See also* Investments
Belgium
  patents, 237n
Bibliographies, 8n, 29n, 43n, 47n, 48n, 65n, 67n, 75n, 129n, 212n
  chemical, 13n
  databases, 235-48
  energy, 3, 4n, 6n, 24n, 78n
  engineering, 15n
  geology, 1, 5, 36n, 99n
  offshore industry, 2, 7n, 109n, 111n
Britain. *See* United Kingdom
Brokers, 146n
Buyers. *See also* Demand; Suppliers
  Europe, 53n, 137n, 189n
  international, 146n
  offshore industry, 205n
  pipeline, 151n
  United States, 169n

Calculations, 81n, 265n. *See also* Conversion factors
  drilling, 82n, 88n
  engineering, 72n
  pipeline, 116n, 117n
  production, 123n

California
  companies, 164, 179n
Canada, 76, 100, 135
  companies, 33n, 77, 143n, 153n, 173n, 184, 186
  databases, 235n, 237n, 243n, 255n, 272n, 277n, 279
  educational programs, 162-63
  patents, 237n
  production, 49n
  products, 208n
  rig information, 277n
  statistics, 49n, 77, 184n
  pipelines, 115
Catalysts. *See* Chemicals
Cathodic protection, 20n. *See also* Corrosion
Central America. *See* Latin America
Chemicals, 9, 13, 126n. *See also* Chemistry; Petrochemicals industry; Products
  databases, 236n, 237n, 257, 274n, 291n
  oilfield, 11, 121n
  patents, 11n, 12, 13n, 237n
Chemistry, 8n, 10n, 73n, 125. *See also* Chemicals; Geochemistry; Petrochemicals industry; Products
  databases, 236n, 237n, 240n, 241n, 246n
  refining, 127n, 128n
China. *See also* Chinese-language sources
  companies, 195
Chinese-language sources, 93n, 195, 248n
Coal, 4n, 6n, 31n, 220n
  databases, 239n, 240n, 242n, 243n, 270n, 282n
  pipelines, 154n
Coatings, 111n, 114n, 151n. *See also* Chemicals; Corrosion
Colleges. *See* Schools
Colorado. *See also* Rocky Mountain region
  companies, 159n, 174n
Companies. *See also* Consultants; Contractors; *individual countries*
  Canada, 77n
  construction, 153n, 155n, 180n, 184n

directories
  Asian, 59n, 194n, 195n, 196
  Australian, 259n
  Canadian, 173n, 184n, 186
  European, 53n, 58n, 137n, 187n, 189n, 190n, 191n, 192n, 197n
  international, 143-55, 170n, 173n, 175n, 206
  Latin American, 193n
  Middle Eastern, 61n, 194n
  United States, 52, 143n, 156n, 157n, 158n, 159n, 160n, 161n, 164-68, 169n, 170n, 171, 172n, 173n, 174n, 175n, 176n, 177n, 178n, 179n, 180n, 181n, 182n, 183n, 213n
drilling, 63n, 143n, 165-68, 173n, 175n, 181n, 259n, 286n
educational programs, 210n
equipment, 207n
Europe, 56n, 108n
exploration, 170n
geophysical, 170n, 181n, 202n
marketing/management, 52n, 231n
natural gas, 137n, 222n
offshore, 58n, 109n, 149n, 202n, 205n
petrochemical, 153n, 197n
pipeline, 144n, 151n, 154n
products, 198n, 200n, 208n
refining, 155n
shipping, 201n
software, 209n
statistics, 231n, 267n, 269
United States, 63n
Completions. *See* Well completion
Concessions, 108, 256n, 285n
Conferences, 2n, 3n, 7n, 11n, 15n, 20n, 23n, 145n, 239n, 241n, 243n, 245n
Conservation, 4n, 12n, 239n, 240n. *See also* Environmental aspects
Construction, 48n, 116n, 117n, 119n, 154n, 193n, 202n, 203n, 204n, 219n, 222n. *See also* Companies; Contractors; Drilling; Pipelines
Consultants. *See also* Personnel
  United Kingdom, 189n
  United States, 141n, 152n, 164n, 165n, 166n, 167n, 179n, 180n

Consumption, 64n, 190n, 220n. *See also* Demand
  databases, 239n, 270n, 275n, 282n, 283n, 292n, 294n, 298n
  statistics, 31n, 59n, 70n, 77n, 146n, 215n, 217n, 222n, 229n, 230n, 231n, 234n, 275n
Continuous tables, 72. *See also* Calculations
Contractors. *See also* Companies; Consultants
  drilling, 139, 143, 152n, 164-68, 173, 184n, 259n
  geophysical, 169n
  offshore, 56n, 110n, 111n, 191n, 202, 203n
  oilwell servicing, 173
  pipeline, 144, 154
Conversion factors, 24n, 27n, 31n, 33n, 35n, 40n, 47n, 64n, 70n, 72n, 82n, 83n, 96n, 98n, 101n, 116n, 119n, 130n, 189n, 204n, 219n, 222n, 229n. *See also* Tables
Core analysis, 98. *See also* Drilling
Corporations. *See* Companies
Corrosion, 14, 20n, 119n, 140n, 248n. *See also* Coatings
Costs, 62n, 64n
  databases, 238n, 263n, 267n, 268n, 269n, 271n, 282n
  determination, 80n, 86n, 114, 224
  drilling, 268n, 271n
  enhanced recovery, 121n
  pipeline, 114, 223n
  statistics, 215n, 222n, 228n
Crackers. *See* Processing; Refineries; Refining
Crude oil, 31n, 59n, 115, 128n, 148n, 154n, 155n, 164n, 166n, 168n, 169n, 183n, 215n, 226n. *See also* Heavy oils
  databases, 234n, 251n, 255n, 263n, 265, 268n, 269n, 270n, 275n, 278, 280n, 282n, 283n, 287, 289n, 290n, 291, 293n, 294n, 298n, 299n
  prices, 64n, 214n, 223n, 229n, 233n, 251n, 263n, 269n, 270n, 275n, 280n, 282n, 283n, 287, 289n, 290n, 291

Crude oil (*continued*)
  statistics, 214n, 223n, 229n, 231n, 233n, 270n, 282n, 294n
Czechoslovakian-language sources, 248n

Danish-language sources, 43n. *See also* Denmark; Scandinavia
Data sheets, 97
Databases
  abstracts, 235n, 236n, 237n, 238n, 240n, 244n, 246n, 248n, 260n
  bibliographic, 235-48
  chemical, 236n, 237n, 240n, 241n, 246n, 257, 274n, 291n
  data only, 262-301
  data/textual, 249-59
  energy, 238n, 239-41, 247n, 248n, 270, 273, 274n, 275, 282-83, 298
  engineering, 236n, 237n, 240n, 241n, 276n
  full text, 260-61
  geology, 242n, 243n, 245n, 246n, 248n, 264n
  statistics, 239n, 241n, 247n, 249n, 256, 262-301
Demand, 3n, 275n. *See also* Consumption
  statistics, 52n, 59n, 214n, 215n, 225n, 226n, 228n, 232n, 270n, 283n, 299n
Denmark. *See also* Danish-language sources; Scandinavia
  offshore industry, 54n, 108n, 261n
Development, 7n, 49, 50n, 55n, 57n, 59n, 61n, 73n, 77n, 110n, 122n, 146n. *See also* Enhanced recovery; Exploration; Production
  databases, 244n, 250n, 260n, 268n, 269n, 276n, 285n, 300n
Dictionaries, 47n, 54n, 64n, 66n, 67n, 68n, 70n, 71n, 73n, 74n, 77n, 79n, 88n, 93n, 113n, 124n, 149n, 203n, 204n, 212n, 219n, 222n, 229n
  abbreviations, 24n, 25n, 27n, 29n, 32n, 33
  general, 24-31
  multilingual, 39-45
  specialized, 32-38

Directories, 4n, 5n, 6n, 7n, 50n, 54, 56n, 100n, 107n, 108n, 145, 150n, 162-63, 188n, 215, 218n, 227n
  companies
    Asian, 59n, 194-96
    Australian, 259n
    Canadian, 173, 184, 186
    European, 53n, 58n, 137, 187, 189, 190n, 191n, 192n, 197n
    international, 143-45, 146n, 147-49, 150n, 151-55, 170, 173, 175, 206n
    Latin American, 193n
    Middle Eastern, 61, 194
    United States, 52n, 143, 156-61, 164-81, 182n, 183, 213
  individuals, 132-42
  products and equipment, 197-213
Discovery wells. *See* Drilling; Exploration; Well data
Distribution. *See* Transportation
Downhole, 82n. *See also* Enhanced recovery; Production
Downstream. *See* Refining
Drilling, 51n. *See also* Companies; Contractors; Development; Drilling fluids; Equipment; Exploration; Rigs; Statistics; Well data
  abstracts, 20n, 21n
  databases, 245n, 248n, 249n, 250n, 254, 256n, 258n, 259n, 262n, 268, 271, 275n, 276n, 277, 283n, 284n, 285n, 286, 300n
  dictionaries, 28n, 38n, 44
  handbooks, 72n, 74n, 75n, 83, 85, 88, 95-96, 98n
  manuals, 82, 89, 94
  offshore industry, 20n, 111n, 112n, 114n, 202n, 203, 211n
  texts, 65n, 66n, 67n, 68n, 84, 86-87, 93
Drilling fluids, 89n. *See also* Chemicals; Drilling
  abstracts, 11n, 13n
  handbooks, 82n, 83n, 85n, 86n, 90n, 91n, 92n
  texts, 87n
Dutch-language sources. *See also* Netherlands
  databases, 236n, 237n, 248n
  dictionaries, 41n, 43n

Earth sciences. *See* Geology; Geophysics
East Germany. *See also* German-
    language sources
  patents, 237n
Ecology. *See* Environmental aspects
Economics, 6n, 19n, 22n, 72n, 73n,
    78-79, 124n, 216n, 230n. *See also*
    Financial data; Marketing data;
    Sales data
  databases, 235n, 236n, 237n, 238n,
    239n, 241n, 242n, 244n, 245n,
    247n
  Middle East, 60n, 61n
  pipeline, 115n, 116n, 119n
  United Kingdom, 57n
Education. *See* Schools
Employment, 172n, 227n. *See also* Con-
    sultants; Directories; Personnel;
    Seismic
Encyclopedias, 4n, 5n, 35, 46-48, 50-51,
    64n
Energy alternatives, 4n, 6n, 10n, 21n,
    145n, 228n. *See also* Coal;
    Energy information; Geothermal
    energy; Hydroelectric power;
    Nuclear energy; Oil shale industry
  databases, 239n, 248n
Energy information. *See also* Statistics
  abstracts, 16, 17n
  atlases, 3n, 31
  bibliographies, 3, 4n, 6n, 24n, 78n
  conferences, 145
  databases, 238n, 239-41, 247n, 270,
    273, 274n, 275, 282-83, 298
  dictionaries, 24, 30
  guides, 4, 6, 8n
  handbooks, 64, 70n, 78, 79n
  yearbooks, 63
Engineering, 76n. *See also* Consultants;
    Personnel; Reservoir engineering
  abstracts, 10n, 15, 18, 20n
  databases, 236n, 237n, 240n, 241n,
    276n
  dictionaries, 41n
  directories, 155n, 184n, 190n
  drilling, 84-85, 86n, 91, 92n
  guides, 8, 76n
  handbooks, 72, 75, 76n, 82n, 91
  offshore industry, 7, 116n
  pipeline, 116n, 117, 151n

  refining, 127, 155n
  schools, 150
  tables, 72
Enhanced recovery, 11n, 13n, 71n, 121,
    240n, 244, 248n, 255n. *See also*
    Development
Environmental aspects
  abstracts, 21n
  bibliographies, 3n
  databases, 235n, 236n, 237n, 239n,
    240n, 244n
  guides, 8n
  handbooks, 64n, 73n
  indexes, 22n
  North Sea, 109
  United Kingdom, 57n
EOR. *See* Enhanced recovery
Equations. *See* Calculations; Continuous
    tables; Conversion factors
Equipment, 68n. *See also* Contractors;
    Rigs; Services; Suppliers
  dictionaries, 42n, 44
  directories, 137n, 152n, 169n, 176n,
    177n, 179n, 181n, 189n, 190n,
    198, 202, 205-7
  drilling, 44, 82n, 83n, 87n, 89n, 90,
    112n
  handbooks, 123n
  offshore industry, 20n, 32n, 56n,
    112n, 182n, 202, 205, 211n
  oilfield, 198, 206
  pipeline, 114n, 118n, 151
  refining, 127n
  valves, 118n
  wellhead, 71n
Europe. *See also* Companies; *individual
    countries*
  databases, 243n, 251n, 257n, 261n,
    263n, 277n, 280n, 291n
  directories, 53-54, 153n, 187-88, 189n,
    190-92, 208n
  information centers, 188
  offshore industry, 54, 58
  yearbooks, 53-54, 55n, 56n, 58
Expenditures. *See* Costs
Exploration, 175n. *See also* Companies;
    Development; Production
  abstracts, 19n, 21n
  Australia, 258n, 259n
  Canada, 49n, 77n, 184n

Exploration (*continued*)
  databases, 242n, 245n, 246n, 248n, 249, 250n, 258n, 259n, 261n, 264n, 266n, 268n, 269n, 270n, 276n, 277n, 283n, 285, 286n, 300n
  dictionaries, 37n, 38n, 41n
  encyclopedias, 51n
  France, 56n, 190n, 261n, 264n
  handbooks, 64n, 73n
  Japan, 59n
  North Sea, 109n, 191n
  Norway, 55n
  offshore, 32n, 56n, 58n, 204n, 261n
  rig information, 259n, 277n
  Scandinavia, 245n
  statistics, 59n, 62n, 70n, 214n, 217n, 225n, 227n, 228n, 268n, 269n, 270n, 283n
  texts, 65n, 66n, 67n, 99n, 122n
  United Kingdom, 57n, 261n
  United States, 49n, 62n, 178n, 179n, 181n, 228n
  well data, 249, 250n, 258n, 276n, 285, 300n
  yearbooks, 49n, 53n, 56n, 59n, 61n, 62n, 63n, 146n
Exploration geophysics. *See* Geophysics
Exports, 59n, 77n, 214n, 231n, 270n, 282n, 283n, 294n, 298n

Facilities, 114n, 122n, 127n
Fact sheets, 55
Field activities. *See* Gas fields; Oilfields
Field code master list, 69
Field courses, 163n. *See also* Schools
Field trip guidebooks. *See* Geology
Financial data, 6n, 52n, 60n, 63n, 206n, 222n, 267n, 274n, 290n. *See also* Costs; Economics; Pricing
Florida
  companies, 158n, 165n, 171n
Flow charts, 38n
Fluids. *See* Drilling fluids; Stimulation fluids
Forecasts, 6n, 80n
  databases, 249n, 252n, 253n, 290n
  future prices, 290n
  petrochemical information, 252n, 253n
  statistics, 215n, 228n
  well yield, 249n
Foreign-language references. *See also individual languages*
  abstracts, 13n
  bibliographies, 1n, 21n, 236n, 238n, 239n, 242n, 243n, 245n, 246n, 247n, 248n
  databases, 236n, 238n, 239n, 242n, 243n, 245n, 246n, 247n, 248n, 264n
  dictionaries, 39-45, 47n
  drilling, 93n
  information centers, 188n
  statistics, 232n
  yearbooks, 53n, 56n
Fossil fuels. *See* Coal; Crude oil; Energy information; Natural gas
Four Corners region, 167. *See also* Arizona; Colorado; New Mexico; Rocky Mountain region; Utah
France. *See also* Europe; French-language sources
  directories, 190n
  exploration data, 190n, 261n, 264n
  offshore industry, 54n, 56, 261n
  patents, 237n
Freight rates, 225n, 229n, 263n, 291n. *See also* Shipping and storage; Tanker fleets; Transportation
French-language sources, 83n. *See also* Europe; France
  databases, 236n, 237n, 238n, 243n, 246n, 247n, 248n, 264n
  dictionaries, 40n, 41n, 43n, 45
  directories, 190n
  encyclopedias, 48n
  statistics, 232n
  yearbooks, 53n, 56
Fuel oil, 220n, 251n, 263n, 298n

Gas, 17, 48, 82, 123n, 137. *See also* Gasoline; Natural gas
Gas fields, 31n, 50n, 69, 187n
Gas processing. *See* Natural gas; Processing
Gas utility industry, 222n

Gasoline, 128n, 265n
  marketing, 251n, 292n, 296n, 297n, 301n
  price information, 251n, 263n, 279, 288n, 291n, 292n, 295n, 296n, 301n
  statistics, 220n, 270n
  volume, 279
Geochemistry, 5n, 21n, 46n, 242n, 243n, 246n, 248n. *See also* Chemistry; Geology
Geology, 76n. *See also* Geophysics
  abstracts, 21n, 235n
  bibliographies, 1, 5, 242n, 243n, 246n
  data sheets, 97
  databases, 235n, 242n, 243n, 244n, 245n, 246n, 248n, 249n, 250n, 264n, 285n
  dictionaries, 28n, 34, 36, 39, 41n
  directories, 132, 135, 138, 141, 189
  educational programs, 163n, 189
  encyclopedias, 46n
  Europe, 54n
  field trip guidebooks, 100
  guides, 5n, 8n
  heavy oils, 244n
  indexes, 1
  North Sea, 110n
  software, 199
  style guides, 131n
  texts, 65n, 66n, 68n, 99
  time scale, 29n, 31n, 35n, 66n, 98n
  United Kingdom, 189n
  United States, 96n, 249n, 250n
  well-site, 101
  yearbooks, 49n
Geophysics. *See also* Companies; Contractors; Geology
  abstracts, 21n
  databases, 243n, 245n, 246n, 248n, 266n
  dictionaries, 28n, 29n, 35
  directories, 162, 169n, 170, 181n, 202n
  educational programs, 162
  statistics, 49n, 225n
Georgia
  companies, 158n, 165
  products, 208n
Geosciences. *See* Geology

Geothermal energy, 4n, 31n, 64n. *See also* Energy alternatives
German-language sources. *See also* East Germany; West Germany
  databases, 236n, 237n, 239n, 248n, 261n
  dictionaries, 39n, 41n, 43n
  yearbooks, 53n
Glossaries. *See* Dictionaries
Government agencies, 22n, 241n. *See also* Law; Lease operations; Legislation; Licensing; Politics
  Canadian, 184n, 235n, 272n
  French, 56n
  Japanese, 59n
  North Sea, 191n
  Norwegian, 103n, 105n
  United Kingdom, 58n, 174n, 189n
  United States, 181n, 183n, 227n, 233n, 272n
Great Britain. *See* United Kingdom
Greece, 54n
Guidebooks. *See* Geology
Guides, 4-6, 8, 76, 218, 223
Gulf Coast. *See also* Alabama; Florida; Louisiana; Mississippi; Texas
  companies, 158, 171
  databases, 291n
  products, 208n
Gulf of Mexico, 254n

Handbooks, 4n, 6n, 64, 70n, 72n, 73-75, 222n, 231n. *See also* Manuals; Texts
  business, 77-79, 80n
  drilling, 83, 85, 88, 91, 92n, 95n, 96
  exploration, 98, 101
  legal, 104, 107
  offshore industry, 111n, 112
  pipeline, 116-20
  production, 123, 231n
  refining, 126
Heavy oils, 31n, 50n, 220n. *See also* Oil sands; Tar sands
  databases, 244, 251n, 263n, 298n
Holland. *See* Dutch-language sources; Netherlands
Hungarian-language sources, 248n

186 / SUBJECT INDEX

Hydrocarbons. *See* Crude oil; Natural gas; Petrochemicals industry
Hydroelectric power, 6n, 31n. *See also* Energy alternatives
Hydrology. *See* Geology

IEA. *See* International Energy Agency
Imports
 databases, 270n, 278, 282n, 283n, 294n, 298n, 299n
 Japanese, 59n
 statistics, 70n, 214n, 217n, 225n, 270n, 282n, 283n, 294n
 United States, 70n, 270n, 278n, 298n, 299n
Indexes, 3, 4n, 22, 69n. *See also* Abstracts
 chemical, 13n
 databases, 235, 236n, 237n, 239, 240n, 242n, 243n, 244, 247n
 engineering, 15, 18, 23
 geological, 1
Indexing services. *See* Indexes
Instrumentation, 68n, 122n, 212n. *See also* Equipment
International Energy Agency, 232n, 294n
Inventories, 270n, 281n, 282n, 283n, 287n, 294n, 298n, 299n. *See also* Storage
Investment banking reports. *See* Banking
Investments, 238n, 268n. *See also* Buyers; Companies; Costs; Lease operations
Ireland
 offshore industry, 54n, 108n, 261n
Italian-language sources, 41n, 43n, 243n. *See also* Italy
Italy, 54n, 289n. *See also* Italian-language sources

Japan, 59. *See also* Japanese-language sources
Japanese-language sources, 236n, 237n, 243n, 248n. *See also* Japan
Journals. *See* Periodicals

Kansas
 companies, 159n, 161, 174n
 products, 208n

Land use, 51n, 67n, 104n, 176n, 181n. *See also* Lease operations; Licensing
Landmen, 37n, 51. *See also* Land use; Lease operations
Latin America
 companies, 153n, 193
 databases, 277n
 products, 208n
 rig information, 277n
Law, 76n, 104n, 107n. *See also* Government agencies; Legislation; Licensing
 dictionaries, 37n
 Norway, 103n, 105n
 property leasing, 51n
 tax, 102n
 United Kingdom, 106
Lease operations, 49n, 66n, 102n, 104n, 176n, 219n, 255n. *See also* Development; Exploration; Government agencies; Land use; Law; Legislation
 forms, 51n, 74n
Legislation, 76n, 104n, 105n, 238n. *See also* Government agencies; Law; Lease operations; Licensing
Licensing. *See also* Government agencies; Law; Lease operations; Legislation
 Europe, 54n, 58n
 Norway, 55n
 United Kingdom, 58n, 106n
Liquefied petroleum gases (LPG), 229n, 257n, 281, 294n
Logging. *See* Well logging
Louisiana
 companies, 158n, 165, 171n, 182n
 drilling, 254n
 products, 208n
LPG. *See* Liquefied petroleum gases

Manuals, 4n, 37. *See also* Handbooks; Texts
  drilling, 82, 86, 88-90, 94
  pipeline, 114
  refining, 129n
  style, 130n, 131n
Maps, 5, 61n, 76n, 107n, 147n, 194n, 227n
  databases, 239n, 242n, 243n
  energy sources, 3n, 31n
  offshore production, 32n, 54n, 57n, 58n, 108n, 109n
  oilfield, 31n, 50n, 53n, 54n, 58n, 108n, 227n
  pipeline, 50n, 53n, 108n, 115n
  refineries, 50n, 53n
  statistics, 38n, 62n, 77n, 230n
  symbols, 29n, 33n
Marketing data, 6n, 68n, 207n. *See also* Economics; Financial data; Pricing
  Canada, 77n
  databases, 251-52, 257n, 270n, 274, 275n, 280n, 288n, 290n, 292, 296-97
  handbooks, 65n, 73n, 74n, 77n
  Japan, 59n
  Norway, 55n
  reports, 251-52, 274
  statistics, 52n, 215n, 231n, 270n
Marketing firms. *See* Companies; Marketing data; Pricing; Sales data
Mathematical symbols, 33n
Mediterranean, 251n, 263n. *See also individual countries*
Meetings. *See* Abstracts; Conferences
Metric equivalencies. *See* Calculations; Conversion factors
Midcontinent region, 174. *See also individual states*
Middle East
  companies, 61n, 153n, 194
  databases, 277n
  yearbooks, 60n, 61n
Mississippi
  companies, 158n, 165, 171n
  products, 208n
Missouri
  companies, 174n

Montana
  companies, 159n
Mooring systems, 20n, 114n. *See also* Equipment; Offshore industry; Rigs
Mud systems. *See* Drilling fluids
Multilingual references. *See* Foreign-language references

Natural gas. *See also* Companies
  databases, 267, 268n, 269n, 282n, 283n, 289n, 294n, 298n
  guides, 4n, 6n
  handbooks, 73n, 78n
  processing, 155, 184n
  statistics, 214n, 215n, 220n, 221-22, 228n, 267, 282n, 283n, 289n, 294n
Nebraska
  companies, 159n, 174n
Netherlands. *See also* Dutch-language sources
  databases, 237n, 261n
  offshore industry, 54n, 108n, 261n
  patents, 237n
New Mexico
  companies, 159n, 160, 167, 182n
New Zealand, 258n
Newsletters, 6n, 253, 261n
North America. *See* Canada; United States
North Carolina
  companies, 158n
North Dakota
  companies, 159n
North Sea, 108-10, 191. *See also individual countries*
Norway, 105. *See also* North Sea; Norwegian-language sources
  databases, 245n, 260n, 261n
  legislation, 103n, 105
  offshore industry, 54n, 55, 103n, 108n
Norwegian-language sources, 103, 245n, 248n. *See also* Norway
Nuclear energy, 4n, 6n, 64n, 240n. *See also* Energy alternatives
  statistics, 220n, 270n

188 / SUBJECT INDEX

OECD. *See* Organization for Economic Co-operation and Development
Offshore industry, 69n. *See also* Companies; Contractors; Drilling; Exploration; Licensing; Maps; Personnel; Pipelines; Rigs; Services; Shipping and storage; Tanker fleets; *individual countries*
   abstracts, 20
   bibliographies, 2, 7
   databases, 245n, 255n, 256n, 260n, 261n, 272n, 277n, 283n, 286n
   dictionaries, 28n, 32, 41n, 43, 54
   directories, 54, 56, 58, 108, 110n, 152n, 165n, 195, 201n, 202-5, 211
   handbooks, 111-12, 116n
   yearbooks, 54, 55n, 56, 58
Oil sands, 78n, 235. *See also* Heavy oils; Tar sands
Oil shale industry, 64n, 78n. *See also* Energy alternatives
Oil spills. *See* Pollution
Oilfield chemicals. *See* Chemicals; Oilfields
Oilfields, 19n, 31n, 66, 69, 108n, 109n, 110n, 121n. *See also* Development; Exploration; Maps
   databases, 249n, 255n, 260n, 284n, 285n, 293n
   dictionaries, 24n, 25n, 41, 88n
   directories, 152n, 161n, 180n, 187n, 198, 206n, 213
   yearbooks, 49n, 50, 53n, 54n, 55n, 58n
Oklahoma
   companies, 159n, 161, 166, 174n, 178, 182n
   databases, 254n
Online formats. *See* Databases
Online services, 4n, 5n, 6n. *See also* Databases
OPEC. *See* Organization of the Petroleum Exporting Countries
Organization for Economic Co-operation and Development, 232n, 294n
Organization of the Petroleum Exporting Countries, 214n, 230, 289n

Organizations. *See* Associations; Companies; Directories

Pacific Coast, 179
Papua New Guinea, 258n
Patents. *See also individual countries*
   abstracts, 11n, 12, 13n, 17n, 20n, 21n
   chemicals, 11n, 12, 13n, 237n
   databases, 237n, 240n, 245n, 248n
   European offices, 188n
   refining, 12, 237n
People's Republic of China. *See* China; Chinese-language sources
Periodicals, 302-420
Permian Basin, 180, 208n. *See also individual states*
Permit information, 262n, 271n, 284. *See also* Government agencies; Lease operations; Licensing
Personnel, 138n, 222n. *See also* Consultants; Contractors
   directories, 53n, 63n, 132-42, 143n, 148n, 149n, 151n, 152n, 153n, 154, 155n, 156n, 157n, 158n, 159n, 160n, 164-68, 169n, 170n, 171n, 172n, 173, 174n, 175n, 176n, 177n, 178n, 180n, 181n, 182n, 183n, 184n, 186n, 187n, 190n, 191n, 192n, 193n, 194n, 195n, 196n, 201n, 202n, 205n, 206n, 213
   geophysicists, 170n
   offshore industry, 191n, 201n, 202n, 205n
   pipelines, 144n
Petrochemicals industry, 30, 32n, 68n, 76n. *See also* Chemicals; Products; Refining
   abstracts, 9n, 10n, 12n, 13n, 15n
   databases, 236n, 237n, 240n, 245n, 248n, 252-53, 257n
   directories, 63n, 153, 184n, 193n, 197
   handbooks, 73n
   patents, 12n
   statistics, 193n, 228n
Physical properties, 48n, 83n, 127n, 129n

Physics, 10n, 240n. *See also* Geophysics
Pipelines, 76n. *See also* Companies; Contractors; Costs; Equipment; Personnel; Statistics; Transportation
  abstracts, 14n, 15n, 20n, 21n
  databases, 236n, 245n, 248n, 267n
  directories, 53n, 144, 146n, 151, 152n, 154, 169n, 180n, 181n, 184n
  handbooks, 70n, 116-20
  manuals, 114
  maps, 50n, 53n, 108n, 115
  offshore industry, 108n, 116
  texts, 65n
Platforms. *See* Drilling; Offshore industry; Rigs
Polish-language sources, 243n, 248n
Politics. *See also* Government agencies; Law; Lease operations; Legislation
  databases, 239n, 245n, 247n
Pollution, 9n, 21n, 103n, 219n, 240n, 246n, 248n. *See also* Environmental aspects
Polymers. *See* Chemicals; Petrochemicals industry
Portugal, 54n. *See also* Portuguese-language sources
Portuguese-language sources, 248n. *See also* Portugal
Pricing. *See also* Buyers; Costs; Crude oil; Gasoline; Marketing data; Sales data
  databases, 251n, 253n, 263n, 269n, 273, 274n, 275n, 279n, 280n, 282n, 283n, 287n, 288n, 289n, 290n, 291-92, 295, 301
  handbooks, 64n, 79n
  manuals, 80n
  products, 200n, 251n, 253n, 270n, 273, 274n, 279n, 280n, 287, 288n, 289n, 290n, 291-92, 295, 301
  software, 199n, 209n
  statistics, 70n, 77n, 214n, 215n, 222n, 223n, 225n, 227n, 229n, 233n, 283n
  training, 210n
Primers. *See* Texts

Processing, 9n, 13n, 128n. *See also* Natural Gas; Refineries; Refining
  dictionaries, 28n
  directories, 153n, 155, 181n, 184n, 197, 200
  handbooks, 73n, 78n, 126
Production. *See also* Companies; Processing; Products
  abstracts, 11n, 14n, 20n, 21n
  Canada, 49n, 77n, 184n, 186n, 255n
  databases, 239n, 242n, 248n, 255n, 261n, 262n, 268n, 269n, 270n, 272n, 276n, 282n, 283n, 285n, 289n, 293, 294n, 298n, 299n
  dictionaries, 28n, 32n, 37n, 38n, 41n
  directories, 148n, 161n, 164-68, 174n, 175n, 178n, 181n, 184n, 186n, 187n, 190n, 191n, 193n, 208n
  Europe, 53n, 58n, 187n
  handbooks, 64n, 70n, 72n, 73n, 74n, 75n, 77n, 78n, 83n, 88n, 107n, 123, 222n, 231n
  North Sea, 108n, 110n, 191n
  Norway, 55n
  offshore, 20n, 32n, 55n, 58n, 108n, 219n
  statistics, 49n, 50n, 64n, 70n, 77n, 83n, 182n, 187n, 194n, 214n, 215n, 217n, 219n, 220n, 222n, 223n, 225n, 226n, 227n, 228n, 229n, 231n, 234n, 272n, 282n, 283n, 294n, 299n
  texts, 65n, 66n, 67n, 68n, 71, 122
  United Kingdom, 57n, 58n
  United States, 49n, 62n, 228n, 272n, 299n
  yearbooks, 49n, 53n, 55n, 58n, 61n, 62n, 146n, 226n, 234n
Products. *See also* Gasoline; Petrochemicals industry; Processing; Refining; Suppliers
  abstracts, 12n, 13n, 15n, 19n
  databases, 235n, 236n, 237n, 251n, 253n, 257n, 263n, 270n, 273n, 274n, 278, 280n, 281n, 287, 289n, 292n, 298n
  dictionaries, 38n
  directories, 152n, 153n, 154n, 169n, 171n, 192n, 195n, 197n, 198n, 200n, 206n, 207n, 208n

Products (*continued*)
  handbooks, 79n, 81n, 92n, 119n
  offshore industry, 58n
  pricing, 251n, 263n, 273n, 280n, 282n, 289n, 292n
  yearbooks, 55n, 58n
Properties. *See* Physical properties
Property evaluation, 80. *See also* Land use; Landmen; Lease operations
Publishers, 2n, 6n, 7n, 199n, 218n. *See also* pp. 158-65

Recovery methods. *See* Enhanced recovery; Heavy oils; Production
Refineries, 3n, 50n, 53n, 115. *See also* Processing; Refining
  databases, 239n, 270n, 275n, 281n, 294n, 298n, 299n
  directories, 152n, 169n, 180n, 181n, 190n
Refining. *See also* Gasoline; Petrochemicals industry; Processing; Refineries
  abstracts, 12n, 13n, 14n, 15n, 19n
  databases, 236n, 237n, 245n, 257n, 283n
  dictionaries, 28n, 32n, 38n
  directories, 53n, 155, 184n, 190n, 193n
  handbooks, 70n, 73n, 77n, 79n, 126n, 129
  patents, 12n, 237n
  statistics, 50n, 59n, 70n, 115n, 146n, 193n, 214n, 225n, 228n, 229n, 231n, 234n, 283n
  texts, 65n, 66n, 68n, 125n, 127-28
  yearbooks, 59n, 61n
Regulatory commissions, 191n. *See also* Government agencies; Licensing
Reliability data, 103n, 112
Remote operating vessels (ROV), 111n, 202n. *See also* Equipment; Offshore industry; Subsea vehicles
Reserves
  analysis, 80n
  Canada, 77n
  databases, 238n, 247n, 248n, 267n, 268n, 269n, 270n, 275n, 283n
  dictionaries, 38n
  handbooks, 64n, 70n, 77n

  North Sea, 110n
  offshore, 110n, 219n
  statistics, 31n, 52n, 70n, 77n, 214n, 215n, 217n, 219n, 220n, 222n, 225n, 233n, 267n, 268n, 269n, 270n, 275n, 283n
  United Kingdom, 57n
  yearbooks, 53n, 57n, 60n
Reservoir engineering, 15n, 21n, 72n, 123n, 248n. *See also* Drilling; Engineering; Reservoir stimulation; Reservoirs
Reservoir stimulation, 71n, 121n, 248n. *See also* Enhanced recovery; Reservoir engineering; Reservoirs
Reservoirs, 65n, 248n, 255n, 293n. *See also* Gas fields; Geology; Oilfields; Reservoir engineering; Reservoir stimulation
Rigs, 20n, 85n, 86n, 87n, 93n, 110, 112n, 114n, 124n, 139n, 202n, 203, 211n. *See also* Drilling; Equipment; Offshore industry; Services
  databases, 248n, 256, 259, 268n, 277, 286n
  statistics, 111n, 203, 223n, 225n, 226n, 256, 268n, 277
Rocky Mountain region, 159, 167, 181. *See also individual states*
ROVs. *See* Remote operating vessels
Russian-language sources, 236n, 237n, 243n, 248n

Sales data. *See also* Buyers; Financial data; Lease operations; Marketing data; Pricing
  gasoline, 292n, 296n, 297n
  natural gas, 267n, 282n
  products, 231n, 292n
  statistics, 59n, 222n, 225n, 231n
Scandinavia, 245n. *See also* North Sea; *individual countries*
Schools, 47n, 150, 162n, 163n, 172n, 189n. *See also* Training
Secretarial guides, 33n, 74
Seismic, 39n, 259n. *See also* Exploration; Geophysics; Personnel
  crews, 266, 268n

Seismology, 28n
Serbo-Croatian-language sources, 248n
Services, 17n, 33n, 100n, 118n. *See also* Companies; Construction; Consultants; Contractors; Drilling; Well servicing
 Asia, 195n, 196n
 directories, 53n, 137n, 142n, 151n, 152n, 153n, 156n, 161n, 165n, 169n, 170n, 171n, 172n, 174n, 175n, 177n, 179n, 180n, 181n, 188n, 189n, 191n, 192n, 195n, 196n, 197n, 198, 200n, 201n, 202n, 205-6, 207n, 208n, 213
 Europe, 53n, 56n, 58n, 188n
 North Sea, 191n
 offshore industry, 56n, 58n, 191n, 201n, 202n, 204-5
 texts, 124
 United Kingdom, 58n, 137n, 189n, 192n, 197n
 United States, 213
 yearbooks, 53n
Shipbuilding, 201. *See also* Tanker fleets
Shipping and storage, 10n, 12n, 21n, 190n, 201, 211n, 236n, 237n, 245n, 248n, 274n. *See also* Storage; Tanker fleets; Tanker terminals; Transportation
Societies. *See* Associations
Software, 199, 209. *See also* Databases
Solar energy, 4n, 6n, 31n, 64n, 240n. *See also* Energy alternatives
South Africa
 patents, 237n
South America. *See* Latin America
South Carolina
 companies, 158n
South Dakota
 companies, 159n
Southeast Asia. *See also* Asia
 companies, 196
Spain, 54n, *See also* Spanish-language sources
Spanish-language sources, 41n, 42, 44, 68n, 93n, 242n, 243n, 248n. *See also* Spain
Spills. *See* Pollution
Spot prices. *See* Pricing
Standards, 20n, 96n, 118n, 120n, 188n

Statistics, 49n, 70n, 135, 150n, 214, 220n, 225, 226n, 227n, 228n, 230, 232-34. *See also* Canada; Companies; Consumption; Crude oil; Databases; Demand; Exploration; Imports; Marketing data; Natural gas; Petrochemicals industry; Pricing; Production; Refining; Reserves; Rigs; Sales data; Tables
 abstracts, 10n, 11n, 21n
 construction, 222
 dictionaries, 38
 drilling, 62n, 70n, 214, 217n, 223, 224n, 227n
 energy, 4n, 6n, 31, 63n, 64n, 215n, 216n, 217n, 218, 220n, 225, 230, 233-34
 Europe, 53n, 187n
 gas, 221n, 222n
 guides, 4n, 6n, 218, 223
 historical, 79n, 113n, 214, 215n, 216n, 225, 229n, 232
 Japan, 59n
 Latin America, 193n
 Middle East, 60n, 61n, 230
 offshore industry, 111n, 112n, 203n, 204n, 219, 224n
 pipeline, 70n, 154n, 219n
 tanker, 113n, 229n
 yearbooks, 49n, 50n, 52n, 53n, 59n, 60n, 62n, 63n, 146n, 234
Stimulation fluids, 11n. *See also* Chemicals; Drilling fluids; Enhanced recovery; Reservoir stimulation; Well stimulation
Stocks. *See* Inventories
Storage, 48n, 64n, 68n, 71n, 79n, 146n, 222n, 236n, 237n, 245n, 248n. *See also* Shipping and storage; Transportation
Strathclyde, 192
Structural engineering. *See* Engineering
Style guides, 130, 131n
Submersibles, 211. *See also* Offshore industry; Remote operating vessels; Subsea vehicles
Subsea vehicles, 111n, 211n, 212. *See also* Remote operating vessels

192 / SUBJECT INDEX

Suppliers. *See also* Companies; Contractors; Services
  directories, 137n, 146n, 148n, 164n, 171n, 176n, 177n, 178n, 180n, 186n, 190n, 191n, 196n, 197, 213
  yearbooks, 53n, 54n, 58n
Supply. *See* Consumption; Demand
Sweden, 54n
Symbols. *See* Abbreviations; Maps; Mathematical symbols
Synfuels, 73n, 145n, 220n, 239n. *See also* Energy alternatives; Oil shale industry; Tar sands

Tables, 50n, 61n, 73n, 102n. *See also* Calculations; Conversion factors
  drilling, 83n, 85n, 87n, 89n, 91n, 92n, 94n, 224n
  engineering, 72
  energy, 64n, 217n, 220n
  geology, 39n, 66n, 97n, 98n, 99n
  measurement, 81
  offshore, 32n, 204n
  pipeline, 114n, 118n, 119n, 120n, 151n
  processing, 126n
  production, 123n
  refining, 127n
  statistical, 62n, 79n, 214n, 217n, 223n, 225n, 227n, 230n, 233n, 234n
Tanker fleets, 113, 225n, 229, 291n
Tanker terminals, 50n
Tar sands, 64n. *See also* Energy alternatives; Heavy oils; Oil sands
Tax information, 54n, 80n, 102, 105n
  databases, 269n, 273n
Tennessee
  companies, 158n
Texas
  companies, 158n, 159n, 160-61, 168, 171n, 172, 174n, 177, 182n
  databases, 254n
  products, 208n
  Railroad Commission, 107
Texts, 65n, 66n, 67n, 68n. *See also* Handbooks; Manuals
  business, 80n
  drilling, 84n, 86n, 87n, 93n

  exploration, 99n
  legal, 102n, 103n, 105n, 106n
  production, 71n, 122n, 124n
  refining, 125n, 127n, 128n
Time scale. *See* Geology
Trade associations, 58n, 63n, 169n, 174n, 191n. *See also* Associations
Trade names, 92n, 197n. *See also* Products
Trade shows, 145. *See also* Conferences
Training, 210. *See also* Schools
Transportation. *See also* Pipelines; Shipping and storage; Storage; Tanker fleets; Tanker terminals
  abstracts, 10n, 12n
  databases, 236n, 237n, 245n, 248n, 283n
  dictionaries, 37n, 41n
  directories, 61n, 152n, 190n, 202n, 208n
  handbooks, 64n, 73n, 74n, 75n, 79n, 107n
  statistics, 79n, 214n, 222n, 228n, 283n
  texts, 67n, 68n
Trends. *See* Forecasts

Undersea vehicles. *See* Remote operating vessels; Subsea vehicles
United Kingdom. *See also* Exploration; Government agencies; North Sea; Production; Services
  databases, 237n, 260n, 261n
  directories, 54n, 137n, 189n, 197
  gas industry, 137n
  geological services, 189n
  handbooks, 106n
  licensing, 58, 106n
  maps, 57, 108n
  offshore industry, 54n, 58, 108n, 260n, 261n
  patents, 12n, 13n, 237n, 240n, 248n
  reserves, 57n
  yearbooks, 57-58
United States, 33n, 69n, 76, 100n, 301n. *See also* Associations; Companies; Consultants; Exploration; Geology; Government agencies; Imports; Production; *individual states*

databases, 237n, 243n, 249n, 250n, 255n, 257n, 262n, 266n, 267n, 269n, 270, 272n, 277n, 278n, 280n, 281n, 287n, 288n, 291n, 295n, 296n, 297n, 298, 299n, 301n
directories, 135n, 143n, 144n, 150n, 156n, 157, 158n, 159, 160n, 161n, 162-63, 164n, 165-68, 169n, 170n, 171-72, 173n, 174n, 175n, 176-79, 180n, 181-83, 208n, 213, 215n
drilling, 143n, 173n, 224n, 249n, 250n, 262n, 277n
educational programs, 162-63
energy, 63, 64n, 270, 298n
forecasts, 228
Geological Survey, 131, 243n
handbooks, 64n, 70n, 104n, 121n
leasing, 104n
liquefied petroleum gas, 257n, 281n
maps, 115
marketing data, 280n, 296n, 297n
patents, 12n, 13n, 237n, 240n, 248n
pipelines, 144n, 267n
pricing, 64n, 287n, 288n, 291n, 295n, 301n
statistics, 49n, 62, 70n, 215n, 224n, 226n, 227-28, 233n, 249n, 255n, 267n, 269n, 277n, 299n
tax laws, 102n
yearbooks, 49n, 62-63
Universities. *See* Schools
Upstream, 67n, 146n, 261n
Utah
companies, 159n

Valves. *See* Equipment

Well classification, 29n, 249n, 250n, 276n, 284n, 285n. *See also* Well completion; Well data
Well completion. *See also* Drilling; Well data; Well logging; Well servicing
abstracts, 21n
databases, 248n, 249n, 250n, 262n, 271n, 276n, 286n, 300

fluids, 11n, 92n
offshore, 111n
statistics, 50n, 111n
texts, 71n, 84n, 86n, 87n, 124n
yearbooks, 50n
Well data, 62n. *See also* Drilling; Exploration; Well classification; Well completion; Well logging
databases, 249n, 250n, 262, 271n, 276, 285n, 286n
statistics, 262, 276
Well logging, 21n, 29n, 33n, 47, 85n, 248n, 254n, 256n
Well servicing. *See also* Services; Workover
abstracts, 21n
databases, 248n
directories, 139n, 152n, 169n, 173, 180n, 181n, 184n, 202n
texts, 124
Well-site. *See* Geology
Well stimulation, 124n. *See also* Stimulation fluids
Well testing, 71n, 85n, 262n, 276n. *See also* Well data
Wellhead prices, 77n, 282n. *See also* Marketing data; Pricing
West Germany. *See also* German-language sources
offshore industry, 54n, 108n, 261n
patents, 237n, 248n
Williston Basin, 167
Wind power, 4n. *See also* Energy alternatives
Workover, 124, 169n, 202n
Wyoming
companies, 159n

Yearbooks, 49n, 50n, 52n, 146
Europe, 53-54, 55n, 56, 58
guides, 4n, 6n
Japan, 59
Middle East, 60, 61n
statistics, 226n, 234
United Kingdom, 57n, 58
United States, 52n, 62n, 63